BestMasters

Mit „BestMasters" zeichnet Springer die besten Masterarbeiten aus, die an renommierten Hochschulen in Deutschland, Österreich und der Schweiz entstanden sind. Die mit Höchstnote ausgezeichneten Arbeiten wurden durch Gutachter zur Veröffentlichung empfohlen und behandeln aktuelle Themen aus unterschiedlichen Fachgebieten der Naturwissenschaften, Psychologie, Technik und Wirtschaftswissenschaften. Die Reihe wendet sich an Praktiker und Wissenschaftler gleichermaßen und soll insbesondere auch Nachwuchswissenschaftlern Orientierung geben.

Springer awards **"BestMasters"** to the best master's theses which have been completed at renowned Universities in Germany, Austria, and Switzerland. The studies received highest marks and were recommended for publication by supervisors. They address current issues from various fields of research in natural sciences, psychology, technology, and economics. The series addresses practitioners as well as scientists and, in particular, offers guidance for early stage researchers.

More information about this series at http://www.springer.com/series/13198

Christian Knittl-Frank

A Concise Semisynthesis of Hederagonic Acid

C–H Activation in Natural Product Synthesis

With a Foreword by Univ.-Prof. Dr. Nuno Maulide

Springer Spektrum

Christian Knittl-Frank
Institute of Organic Chemistry
University of Vienna
Vienna, Austria

ISSN 2625-3577   ISSN 2625-3615   (electronic)
BestMasters
ISBN 978-3-658-30010-4   ISBN 978-3-658-30011-1   (eBook)
https://doi.org/10.1007/978-3-658-30011-1

This Springer Spektrum imprint is published by the registered company Springer Fachmedien Wiesbaden GmbH part of Springer Nature.
The registered company address is: Abraham-Lincoln-Str. 46, 65189 Wiesbaden, Germany

Der Zukunft gewidment
Nura und Noah

Foreword

Natural products are chemical compounds that are produced by living organisms. In 1891, Albrecht Kossel suggested a formal division of natural products into two subgroups: primary and secondary metabolites. Primary metabolites – like amino acids, lipids, sugars and nucleobases – are ubiquitous in all living organisms and are essential to life. Secondary metabolites, on the contrary, lack these features; rather, specific secondary metabolites are only produced by specific organisms. The chemical structure of secondary metabolites, the result of millions of years of evolutionary processes, impart evolutionary advantages to the organisms producing them. Some prominent examples of secondary metabolites are well known compounds as caffeine (coffee seeds, tea leaves), nicotine (tobacco leaves) and quinine (bark of the so-called 'chinchona tree'). The fact that living organisms focus on producing only a specific set of natural products is as unique as limiting, especially when considering medicinal applications thereof. Chemists are needed to produce variations of these metabolites – so called analogues – that are not found in nature, as these non-natural analogues might possess enhanced biological activities. Furthermore, many secondary metabolites are produced in minor quantities, which makes their isolation inefficient and expensive, and thus limits commercial availability. Chemists can use so-called 'total synthesis' to produce such molecules from readily available and abundant feedstocks in the laboratory. The author of this book has worked in our team on the development of a new synthetic route towards the natural product hederagonic acid, belonging to the oleanane family. This work forms the core of a bigger endeavor targeting other, higher oxidized members of the oleanane family.

Univ.-Prof. Dr. Nuno Maulide
Head of the Institute of Organic Chemistry

Preface

This book represents the results that were obtained during my master internship in the Maulide Group at the University of Vienna. Certainly, such a work cannot be conducted without the support from several persons, whom I feel obliged to express my sincere appreciation to at this point.

Univ.-Prof. Dr. **Nuno Maulide** for giving me the opportunity to work in his group on this very interesting natural product synthesis.

Dr. **Martin Berger** for supervising my practical work.

The **Maulide Group**, especially lab 1136 (At the time this work was performed: Dr. **Pauline Adler** and Dr. **Rik Oost**) for the very nice working atmosphere. **Martina Drescher** for the exceptional organisation of the labs and equipment. Dr. **Tobias Stopka**, Dr. **Christopher Teskey**, and Dr. **Alexandre Pinto** for proofreading of this thesis.

The **NMR-Team**, the **MS-Team** and Dipl.-Ing. (FH) **Alexander Roller** for measurements of my compounds.

My family, my partner **Mariam Hohagen**, BSc MSc whom without this work would not have been possible, and our childen **Nura** and **Noah** for accompanying me on my ways.

Springer Fachmedien Wiesbaden GmbH for considering my MSc thesis for the BestMasters award and Dr. **Angelika Schulz** for valuable support in the publishing process.

Experimental procedures for **1**, **6b**, **7b** and **8b** as well as spectroscopic data of **7b** and **8b** will be published elsewhere.

Christian Knittl-Frank

Contents

List of Figures

List of Schemes

List of Tables

List of Abbreviations

ACN acetonitrile

d doublet
DCM dichloromethane
DMF dimethyl formamide
DMSO dimethyl sulfoxide

EDTA ethylenediaminetetraacetic acid
ESI electrospray ionization

GBL γ-butyrolactone

HRMS high resolution mass spectrometry

IR infrared

m multiplet

NBS *N*-bromosuccinimide
NCS *N*-chlorosuccinimide

NMR nuclear magnetic resonance

PCC pyridinium chlorochromate

q quartet
quant quantitative
quint quintet

s singlet

t triplet
TCCA trichloroisocyanuric acid
THF tetrahydrofuran
TLC thin layer chromatography
TMS tetramethylsilane
TOF time of flight

UV ultraviolet

X conversion

Y yield

1 Introduction

1.1 Hederagonic Acid

Hederagonic acid (23-hydroxy-3-oxo-olean-12-en-28-oic acid) (**1**) is a naturally occurring pentacyclic triterpenoid based on the oleanane scaffold (Figure 1.1). The molecule is related to hederagenin (**2**), found in *Hedera helix* (common ivy).[6] Hederagonic acid (**1**) can in principle be accessed from hederagenin (**2**) *via* oxidation of the C-3 hydroxy group. The first synthesis of 23-*O*-methyl hederagonic acid methyl ester was reported by Jacobs in 1925.[7] 23-*O*-Methyl hederagonic acid methyl ester was obtained *via* oxidation of 23-*O*-methyl hederagenin with potassium permanganate.[7]

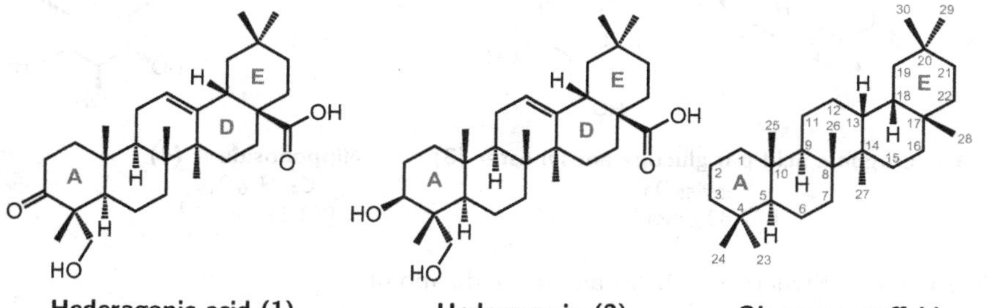

Hederagonic acid (**1**)	Hederagenin (**2**)	Oleanane scaffold
$C_{30}H_{46}O_4$	$C_{30}H_{48}O_4$	$C_{30}H_{52}$
$470.69 \, g \, mol^{-1}$	$472.71 \, g \, mol^{-1}$	$412.75 \, g \, mol^{-1}$

Figure 1.1: Structure of hederagonic acid (**1**), hederagenin (**2**) and the oleanae scaffold with applied atom numbering.

The first reported isolation of hederagonic acid was by Agarwal and Rastogi from *Viburnum erubescens* Wall in 1974.[8] It should be mentioned here, that the use of the therein proposed name 'hederagenic acid' was

© The Editor(s) (if applicable) and The Author(s), under exclusive license to
Springer Fachmedien Wiesbaden GmbH, part of Springer Nature 2020
C. Knittl-Frank, *A Concise Semisynthesis of Hederagonic Acid*, BestMasters,
https://doi.org/10.1007/978-3-658-30011-1_1

widely discontinued, as 'hederagonic acid' had already been coined as trivial name.[9]

Like many other oleananes,[10] hederagonic acid (**1**) is a sapogenin (aglycon). To this day, isolation of two hederagonic acid saponins, hederagonic acid β-D-glucopyranosyl ester (**3**),[11] and nipponoside A (**4**),[12] has been reported (Figure 1.2).

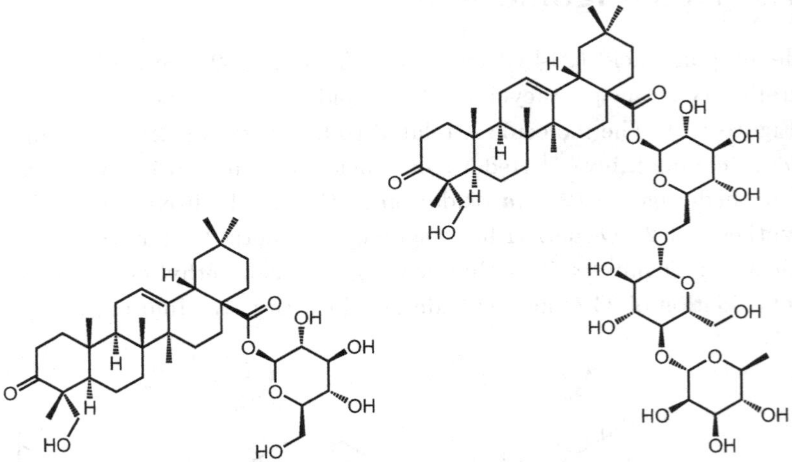

Hederagonic acid β-d-glucopyranosyl ester (3)
$C_{36}H_{56}O_9$
$632.82\,\mathrm{g\,mol^{-1}}$

Nipponoside A (4)
$C_{48}H_{76}O_{18}$
$941.11\,\mathrm{g\,mol^{-1}}$

Figure 1.2: Structure of hederagonic acid saponins.

1.1.1 Natural Occurrence

Hederagonic acid is found in several other plant species besides *Viburnum erubescens*. In 1993, Greca et al. reported its isolation from *Hydrocotyle ranunculoides* (floating pennywort),[13] documented as an invasive species.[14] In 2012, Li et al. detected hederagonic acid as a metabolite in *Celastrus orbiculatus* (oriental bittersweet),[15] native to China,[15] and also an invasive species.[16] In 2012, Yao et al. isolated hederagonic acid (**1**) alongside its saponin hederagonic acid β-D-glucopyranosyl ester (**3**) from *Kalopanax* *septemlobus* (castor aralia),[11] a plant species native to eastern Asia, China, Japan, Korea and eastern coastal Russia.[17] Further species reported to contain hederagonic acid are *Pulsatilla chinensis*,[18] *Pulsatilla cernua*,[19] and *Tripterygium wilfordii* (thunder god vine).[20] Miyakoshi et al. isolated the saponin nipponoside A (**4**) from *Acanthopanax nipponicus*.[12] Selected plant species are depicted in Figure 1.3.

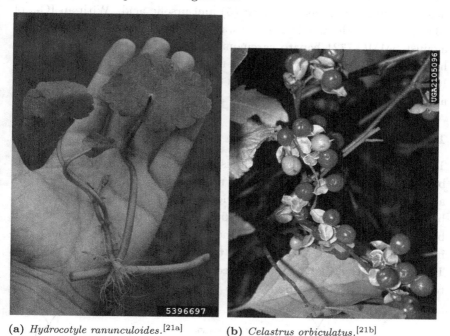

(**a**) *Hydrocotyle ranunculoides*.[21a]   (**b**) *Celastrus orbiculatus*.[21b]

Figure 1.3: Selected plants containing hederagonic acid (**1**).

1.1.2 Biological Activity

It has been reported, that hederagonic acid (**1**) and other triterpenoids arjunolic acid, bayogenin and 4-*epi*-hederagonic acid were moderate glycogen phosphorylase inhibitors.[1,22] Hederagonic acid (**1**) was the most potent inhibitor with an IC_{50} of $54\,\mu mol\,l^{-1}$. Notably, 4-*epi*-hederagonic acid was the second most active inhibitor with an IC_{50} of $62\,\mu mol\,l^{-1}$. The IC_{50}

values of the other examined triterpenoids were $68\,\mu mol\,l^{-1}$ for bayogenin, and $103\,\mu mol\,l^{-1}$ for arjunolic acid.

Furthermore, hederagonic acid (**1**) exhibited moderate cytotoxicity against several cancer cell lines in biological studies.[18] Considering the IC_{50} values, hederagonic acid (**1**) was among the most active compounds when compared with other triterpenoids, such as betulinic acid, hederagenin (**2**), 23-hydroxybetulinic acid, oleanolic acid (**5**) and ursolic acid. With an IC_{50} of $13.02\,\mu g\,ml^{-1}$, hederagonic acid (**1**) was the second most active compound against SMMC-7721 hepato carcinoma cells, and was found to have the highest inhibitory effect against HeLa uterocervical carcinoma cells with an IC_{50} of $11.88\,\mu g\,ml^{-1}$ and against HL-60 leukocythemia carcinoma cells with an IC_{50} of $14.35\,\mu g\,ml^{-1}$.

1.1.3 Previous Syntheses

To date, the first and only semisynthesis of hederagonic acid (**1**) was reported by Wen et al. in 2010.[1] The 9-step synthesis starts from oleanolic acid (**5**), a natural product found in a wide variety of plants.[23] The synthetic route begins with a stepwise benzylation-oxidation-oximation sequence reported by Chen et al.[24] After formation of the benzylester of oleanolic acid, pyridinium chlorochromate (PCC) is used to oxidise the secondary C-3 alcohol. The ketone intermediate is then treated with hydroxylamine hydrochloride to give the corresponding **6c** (Scheme 1.1).

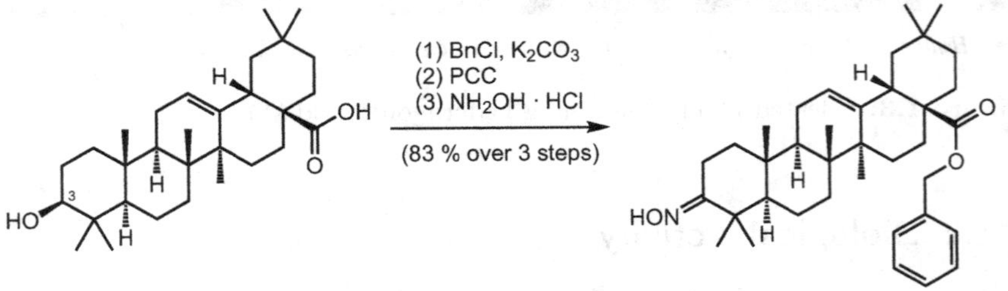

Scheme 1.1: Steps 1–3 of the hederagonic acid (**1**) semisynthesis reported by Wen et al.

Introduction of the C-23 hydroxy group was achieved in 3 steps, utilising Baldwin's stoichiometric cyclopalladation reaction (Scheme 1.2). The intermediate **7c** was obtained as a mixture of diastereomers, which could be separated later in the sequence. The details of this cyclopalladation method are described in Section 1.2.1.

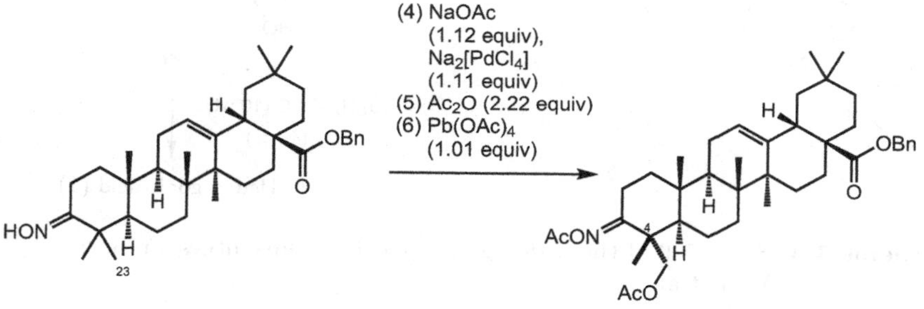

Scheme 1.2: Steps 4–6 of the hederagonic acid (**1**) semisynthesis reported by Wen et al.

The synthesis was finished by a 3-step deprotection series, starting with the deacetylation of O-23 using sodium carbonate in MeOH. At that point the cyclopalladation by-product (In effect a precursor to 4-*epi*-hederagonic acid) was separated *via* column chromatography. Subsequent C-3 deoximation with titanium(III) chloride and water gave ketoalcohol benzylester **8c**, which was finally debenzylated by palladium catalysed hydrogenolysis to yield the target molecule **1** (Scheme 1.3).

A major drawback of this synthesis is the high number of steps combined with the utilisation of stoichiometric amounts of expensive and toxic palladium(II) and lead(IV) compounds for the introduction of the hydroxy group at C-23. Furthermore, the reported synthesis relies on the carcinogenic chromium(VI) reagent PCC for the oxidation of the C-3 hydroxy group. Last but not least, the choice of a benzyl protection resulted in a fair yield for the final deprotection.

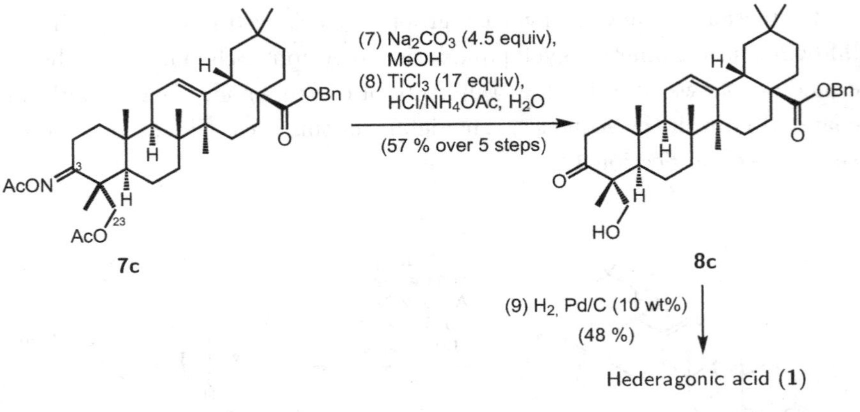

Scheme 1.3: Steps 7–9 of the hederagonic acid (**1**) semisynthesis reported by Wen et al.

1.2 C–H Activation

1.2.1 Stoichiometric C–H Oxidation

1.2.1.1 Baldwin's Cyclopalladation Reaction

In 1985, Baldwin et al. described a method for an oxime directed functional-isation of unactivated primary sp^3 carbons.[25] These authors investigated the reactivity of the organopalladium species **10** and **11**. The dimer **10** was prepared from pinacolone oxime (**9**) via treatment with stoichiometric amounts of NaOAc and sodium tetrachloropalladate(II) in MeOH as de-scribed by Constable et al.[26] Constable et al. have further described the formation of monomeric bridge-split derivatives (**11**). These were obtained via reaction with pyridine, PMe$_2$Ph or PPh$_3$. Baldwin et al. investigated the oxidation of the dimeric species **10** and the monomeric pyridine complex **11a** with Pb(OAc)$_4$.[25] While the former did not react with Pb(OAc)$_4$, the latter was readily oxidised with Pb(OAc)$_4$ at room temperature, giving different products depending on the amount of oxidant used. Their findings indicated that treatment with 1 equiv and 2 equiv of oxidant yielded acet-oxylated oxime **12** and acetoxylated ketone **13**, respectively (Scheme 1.4). To prevent recomplexation of Pd^{II} with products bearing an oxime moiety,

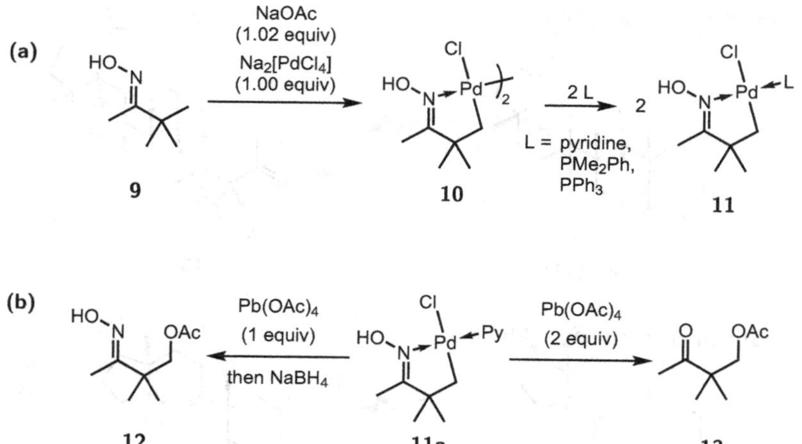

Scheme 1.4: Stoichiometric C–H bond oxidation. (a) The synthesis of monomeric palladacycle **11**. (b) Baldwin's cyclopalladation method.

the Pb(OAc)$_4$ promoted oxidation was followed by reduction of Pd^{II} to Pd^0 with NaBH$_4$.

Baldwin's cyclopalladation reaction has found wide application in the synthesis and modification of natural products and their analogues. Baldwin et al. used it for derivatisation of the triterpene lupanone (**14**) (Scheme 1.5a).[25] Carr et al. used the method to introduce a hydroxy group at C-23 of lanostenone (**16**), enabling 4α-demethylation to obtain 4β-demethyl-lanostenone (**17**) (Scheme 1.5b).[27] Peakman et al. applied Baldwin's method for the introduction of a hydroxy group at C-23 of ursenone oxime, oleanenone oxime and lupanone oxime.[28] Bore et al. have utilised it in the transformation of ursolic acid (**18**) into β-boswellic acid analogues (**19a,b**) (Scheme 1.5c).[29] Finally, as described in Section 1.1.3, Wen et al. have employed this method for introduction of the C-23 hydroxy group in their semisynthesis of hederagonic acid (**1**) (Scheme 1.5d).[1]

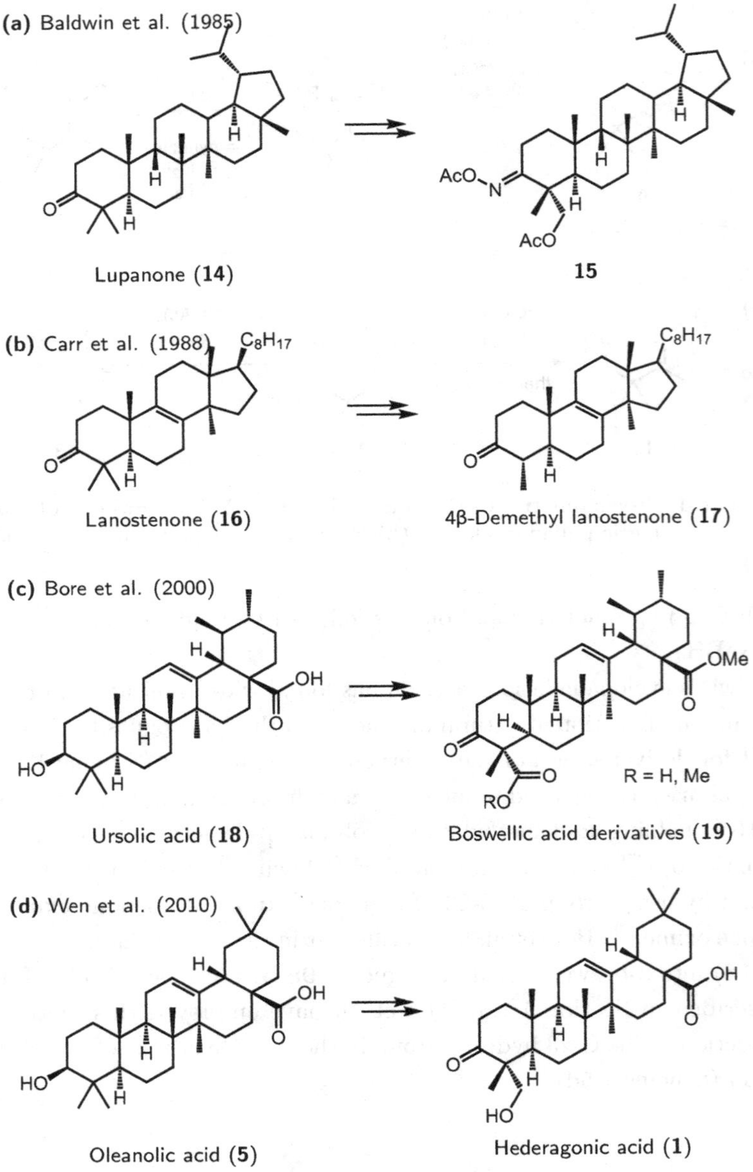
(a) Baldwin et al. (1985)
H
H
O
H
H
H
AcO
N
H
AcO
Lupanone (14)
15
(b) Carr et al. (1988)
C8H17
O
H
C8H17
O
H
Lanostenone (16)
4β-Demethyl lanostenone (17)
(c) Bore et al. (2000)
H
OH
O
H
HO
H
H
OMe
O
H
H
O
O
RO
R = H, Me
Ursolic acid (18)
Boswellic acid derivatives (19)
(d) Wen et al. (2010)
H
OH
O
H
HO
H
H
OH
O
H
O
H
HO
Oleanolic acid (5)
Hederagonic acid (1)

Scheme 1.5: Selected applications of Baldwin's cyclopalladation reaction.

1.2.2 Catalytic C–H Oxidation

1.2.2.1 Palladium Catalysed C–H Acetoxylation

In 2004, Sanford and coworkers reported a modified version of Baldwin's method.[2] This modification utilises catalytic palladium in combination with stoichiometric amounts of an oxidant at elevated temperatures (80 to 100 °C). In the original publication, Sanford and coworkers employed (Diacetoxyiodo)benzene (PhI(OAc)$_2$), however, application of other oxidants such as Oxone has been reported.[30]

In their initial studies, Sanford and coworkers investigated the C–H acetoxylation of pinacolone O-methyl oxime (**20**).[2] They were able to mimic Balwin's method with 5 mol % of Pd(OAc)$_2$ and 1.1 equiv of PhI(OAc)$_2$. However, they obtained a mixture of mono-, di-, and tri-acetoxylated O-methyl oximes **22a–c**. They propose a dimeric intermediate **21**, which was oxidatively cleaved by PhI(OAc)$_2$ to yield the mono-acetoxylated oxime **22a**. Repeated acetoxylation of **22a** lead to di- and tri-acetoxylated oximes **22b,c**.

22	R^1	R^2
a	H	H
b	H	OAc
c	OAc	OAc

Scheme 1.6: Catalytic C–H bond oxidation reported by the the Sanford Group.

These authors further report on the chemo- and regioselectivity of the developed acetoxylation, demonstrating that primary C–H bonds are favoured over secondary C–H bonds. Furthermore, they showed that acetoxylation takes place exclusively in the β-position. This regioselectivity mirrors the highly favoured formation of 5-membered palladacycles.[31] Notably, the successful acetoxylation of 2,2-dimethylpentanone O-methyl oxime (**24**), a

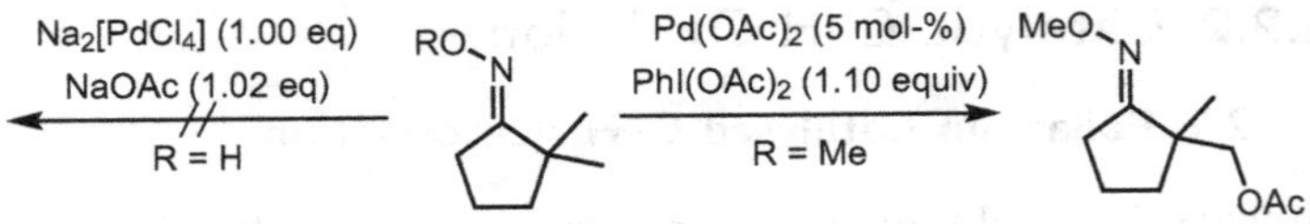

Scheme 1.7: Catalytic C–H bond oxidation of 2,2-dimethylpentanone oximes
23 and **24**.

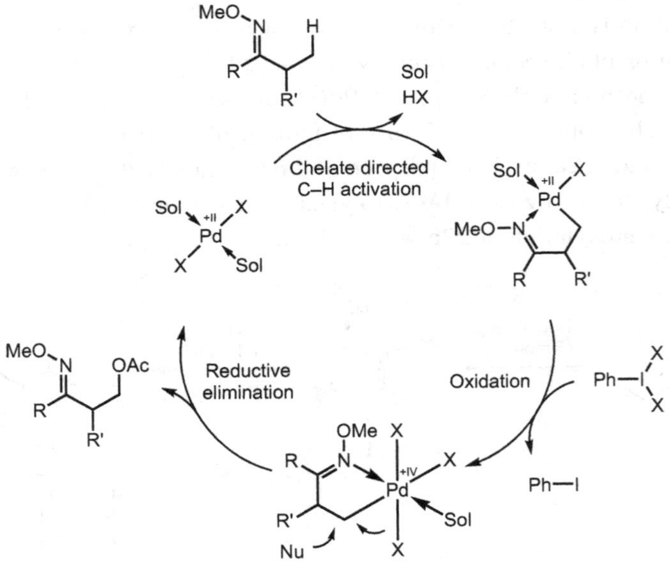

Scheme 1.8: Catalytic cycle of the C–H bond oxidation, proposed by the the
Sanford Group.

molecule that Baldwin et al. have described as unsuited for application of
their method (Scheme 1.7), is reported.[25]

Sanford and coworkers have furthermore conducted mechanistic studies on
this palladium catalysed C–H bond oxidation, and propose a catalytic cycle
featuring a Pd^{II}/Pd^{IV} redox system.[3,32] The catalytic cycle starts with
a chelate-directed C–H bond activation giving a 5-membered palladacycle

intermediate. Next, the oxidation of Pd^{II} to Pd^{IV} is facilitated by the stoichiometric oxidant. The catalyst is regenerated *via* a C–X bond forming reductive elimination, releasing the oxidised product. Both an intramolecular elimination and an intermolecular nucleophilic substitution can facilitate the C–X bond formation.[3]

2 Results and Discussion

2.1 Objectives

Hederagonic acid was found to be a suitable starting material for the synthesis of polyhydroxylated triterpenoids. Nevertheless, hederagonic acid is very expensive and has limited commercial availability. Hederagonic acid can be synthetically accessed from oleanolic acid, a cheap and readily available natural product. However, the to date only reported synthesis of hederagonic acid requires 9 steps, starting from oleanolic acid, and utilises stoichiometric amounts of palladium for the C–H oxidation at position C-23 (Section 1.1.3). Addressing these drawbacks, this work aims to provide a shorter and cheaper synthesis of hederagonic acid to enhance the accessibility of hederagonic acid as a starting material for semisyntheses. Furthermore, synthetic intermediates that can serve as a starting point to access further polyhydroxylated triterpenoids were targeted.

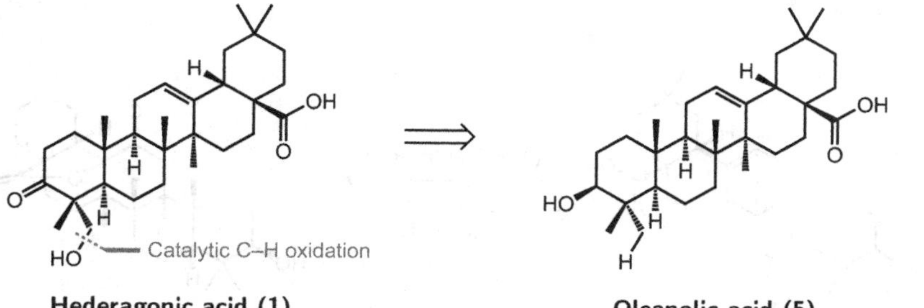

Hederagonic acid (1) **Oleanolic acid (5)**

Scheme 2.1: Concise synthesis of hederagonic acid (**1**) *via* catalytic C–H oxidation.

C. Knittl-Frank, *A Concise Semisynthesis of Hederagonic Acid*, BestMasters,
https://doi.org/10.1007/978-3-658-30011-1_2

2.2 Preliminary Results

Part of the results described in this chapter were obtained in collaboration with Martin Berger, MSc. His preliminary results are summarised in Sections 2.2.1–2.2.4.[33]

2.2.1 Halolactone Oxime

The low yield of the debenzylation step in the synthesis reported by Wen et al. (Section 1.1.3) suggested that an alternative protecting group would be necessary. Moreover, Berger found incompatibilities between the C-12 olefin moiety and reaction conditions needed towards the synthesis of terminolic acid (**33d**). These two problems led to the utilisation of a halolactone group, which can be considered as a 2-in-1 protection of the carboxylic acid and the olefin moiety. Due to an increased C–X bond stability and the consequent robustness through the synthetic sequence,[34] the use of a chlorolactone protection was preferred over a bromolactone protection.

2.2.1.1 Chlorolactone Oxime

The synthesis of chlorolactone oxime **6a**, whose oxime moiety acts as the directing group in the subsequent C–H activation (Section 2.2.2), was initially developed by Berger as a sequential 3-step procedure. The first step involved the use of *N*-chlorosuccinimide (NCS) in dimethyl formamide (DMF) at

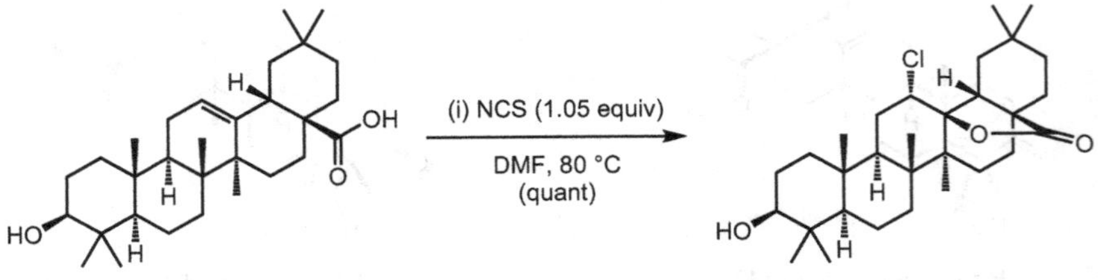

Scheme 2.2: Synthesis of chlorolactone alcohol **26a**. Reagents and conditions: (i) NCS, DMF, 80 °C, 30 min.

80 °C (Scheme 2.2). After aqueous workup, the obtained chlorolactone **26a** was used without further purification.

Chlorolactone ketone **27a** was synthesised *via* oxidation of **26a** with PCC in dichloromethane (DCM) at room temperature (Scheme 2.3). To ensure full conversion at short reaction times 2 equiv of PCC were used. As for the lactonisation, this reaction proceeded with excellent selectivity, so that no further purification was necessary.

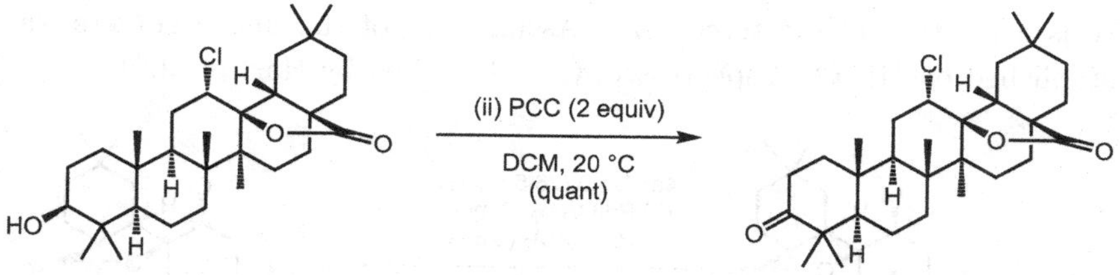

Scheme 2.3: Synthesis of chlorolactone ketone **27a**. Reagents and conditions: (iia) PCC, DCM, 20 °C, 3 h.

Subsequent transformation of chlorolactone ketone **27a** to the respective oxime **6a** was achieved by treatment with 2 equiv of hydroxylamine hydrochloride in pyridine at 80 °C (Scheme 2.4). The obtained oxime was used after aqueous workup without further purification.

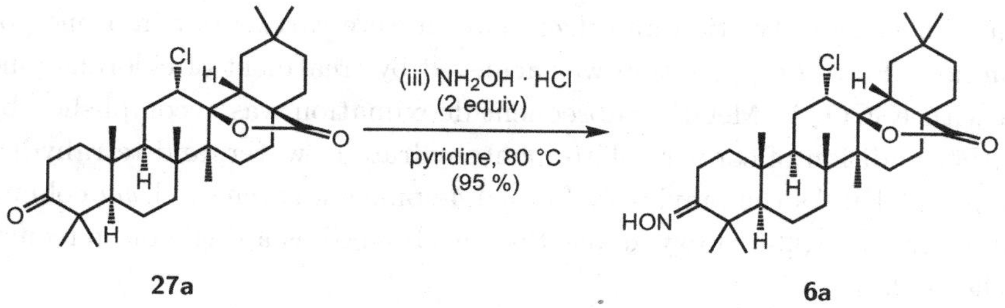

Scheme 2.4: Synthesis of chlorolactone oxime **6a**. Reagents and conditions: (iii) Hydroxylamine hydrochloride, pyridine, 80 °C, 90 min.

2.2.2 C–H Activation at position C-23

2.2.2.1 Palladium Catalysed C–H Acetoxylation

The palladium catalysed C–H activation method, described in Section 1.2.2.1, was utilised for the introduction of the hydroxy group at C-23. The conditions used were based on those reported by Neufeldt and Sanford, using $PhI(OAc)_2$ as the oxidant (Scheme 2.5).[35] The C–H acetoxylation method, with substrate controlled stereoselectivity, yielded the acetoxylated molecule **7a** as a mixture of diastereomers. Assignment of the diastereomers was established by ^{1}H-NMR spectroscopy, as described by Hart et al.[9]

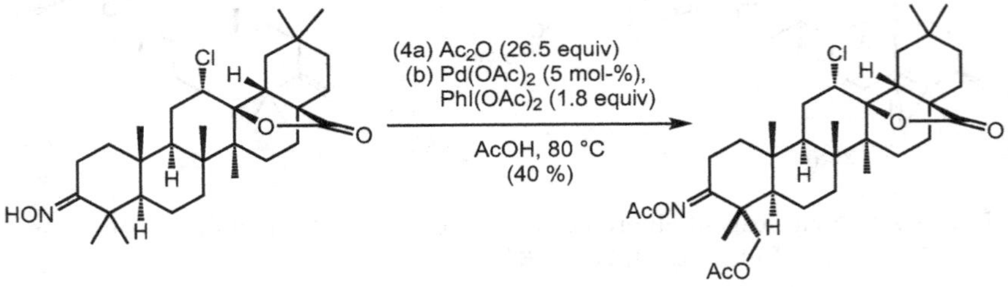

Scheme 2.5: C–H Acetoxylation of chlorolactone oxime **6a** as performed by Berger.

2.2.3 Deacetylation and Deoximation

Subsequent deacetylation and deoximation were carried out in a one-pot reaction. First, deacetylation was achieved by treatment of chlorolactone **7a** with K_2CO_3 in MeOH. Subsequent deoximation was accomplished by further addition of copper sulfate pentahydrate in water, and tetrahydrofuran (THF). As the previously formed 4-epimer was removed *via* column chromatography, β-hydroxy ketone **8a** was obtained as a single diastereomer (Scheme 2.6).

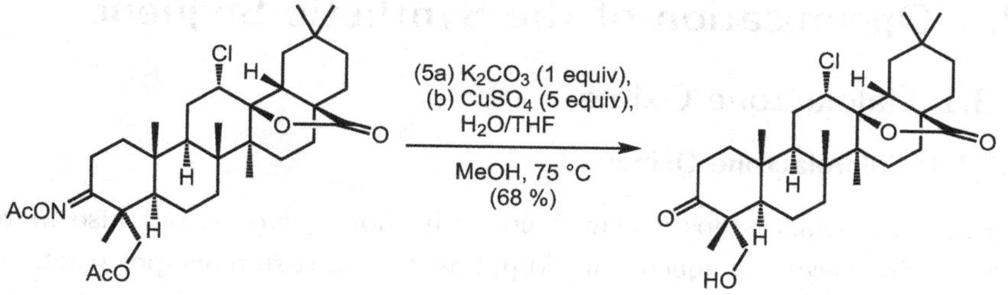

7a
(dr 80:20)

8a
(dr > 20:1)

Scheme 2.6: One-pot deacetylation-deoximation of chlorolactone **7a**. Reagents and conditions: See Scheme 2.25 on Page 37.

2.2.4 Retro-halolactonisation

Lewis and Tucker reported that bromolactone alcohol **26b** could be debrominated to yield oleanolic acid (**5**).[36] The described reaction utilised zinc in acetic acid (Scheme 2.7).

Nevertheless, when Berger applied this method to a more complex 3-oxo-chlorolactone intermediate, it was not successful in opening the lactone. Furthermore, he observed the reduction of the C-3 carbonyl during the reaction. Berger also found that treatment of a more complex intermediate with samarium(II) iodide did not yield the desired free acid.

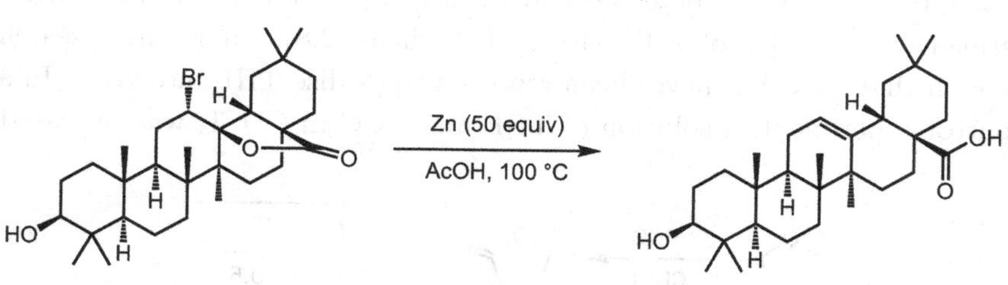

26b

5

Scheme 2.7: Retro-halolactonisation of bromolactone alcohol **26b**, reported by Lewis and Tucker.

2.3 Optimisation of the Synthetic Sequence

2.3.1 Halolactone Oxime

2.3.1.1 Chlorolactone Oxime

As multiple single steps consume not only more solvents, but also more time during necessary aqueous workup, possible multi-step one-pot reactions are preferred.[37,38] To optimise the synthesis of chlorolactone oxime **6a** (Section 2.2.1.1) in terms of step economy, the initial 3-step sequence was examined.

Analysing the first two steps, no conflicting reactivity was anticipated between the chlorination by-product, succinimide, and PCC. Furthermore, oxidation with chromium(VI) in DMF has been reported previously.[39] However, it was found that the oxidation in DMF proceeds much slower. Even when using 3 equiv of oxidant and heating to 80 °C the reaction did not reach full conversion in a less than 24 h. Trace amounts of residual DMF further complicated the removal of chromium(IV) waste-products. When using DCM as solvent, these can normally be removed by precipitation with Et_2O and subsequent filtration through Florisil.[40] Nevertheless, the use of DCM as alternative solvent was not considered as it does not solubilise oleanolic acid (**5**).

During screening of further solvents, THF was found to be a suitable substitute. Not only was oleanolic acid (**5**) soluble, but also the chromium(IV) by-products were easily removable by precipitation with Et_2O. Surprisingly, a new by-product was detected and identified as γ-butyrolactone (GBL), formed *via* oxidation of THF with PCC (Scheme 2.8). Literature research showed that PCC has never been reported to oxidise THF into GBL. In a control experiment, a solution of THF and PCC in $CDCl_3$ was prepared.

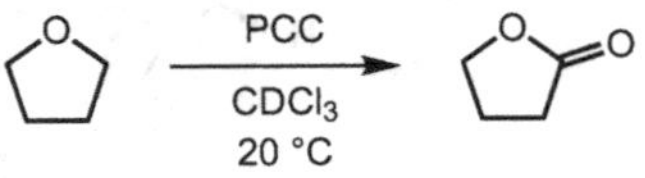

t	$X/\%$
24 h	0.5
12 d	3

PCC, $CDCl_3$, 20 °C

Scheme 2.8: The formation of γ-butyrolactone from THF and PCC.

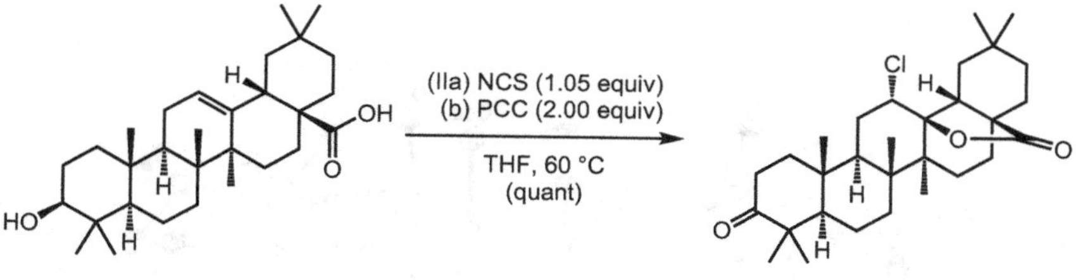

Scheme 2.9: One-pot synthesis of chlorolactone ketone **27a**. Reagents and conditions: (II) THF, 60 °C. (a) NCS, 1 h. (b) PCC, 3 h.

GBL was formed at room temperature, with conversions of 0.5 % after 24 h and 3 % after 12 d.

This side reaction did not, however, pose a limitation as the formed γ-butyrolactone could be readily removed *in vacuo* at 80 °C, without decomposing chlorolactone ketone **27a** (Scheme 2.9). The one-pot sequence gave chlorolactone ketone **27a** in quantitative yield.

Rivero-Chan et al. reported that the bromo-analogue of ketone **27a** (cf. Section 2.3.1.2) readily crystallised during isolation.[41] This was also the case for the herein synthesised chlorolactone ketone **27a**. Dissolution in a 1:1 mixture of DCM and Et$_2$O followed by slow evaporation gave colourless crystals suitable for X-ray analysis. The obtained crystal structure (Figure 2.1) shows that **27a** is a hexacyclic molecule based on the pentacyclic oleanane scaffold. The chlorine atom is tethered to C-12 in an α-axial orientation. The γ-lactone bridges C-17 and C-13 in β-orientation. As expected, the reported bromolactone ketone and the herein synthesised chlorolactone ketone **27a** share the same structural features. Rings A–B, B–C and C–D are *trans*-fused, while rings D–E are *cis*-fused.

After successful combination of the first two steps, a potential concatenation with the oximation step was next investigated (Scheme 2.10). This task however, proved to be more challenging. The lower yield of <86 % was most likely attributable to separation problems during aqueous workup. The latter resulted from problems of the removal of chromium by-products. Separation of the heavy metal by-products worked well in the tandem

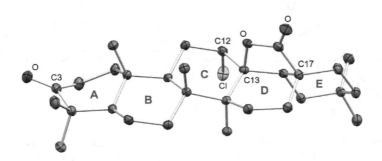

Figure 2.1: Crystal structure of chlorolactone ketone **27a**. Thermal ellipsoids
have been drawn at 50 % probability level. Hydrogen atoms were
omitted for clarity.

lactonisation-oxidation. Thus, it is probable that the presence of hydroxylamine hydrochloride caused the formation of further reduced chromium
compounds. However, attempts to determine the nature of these by-products
were unsuccessful. These newly-produced chromium species were soluble in
the organic phase and could therefore not be removed by filtration.

Numerous attempts were carried out, aiming to remove the chromium
impurities. First assays to separate the heavy metal species during aqueous
workup failed. Neither acidic nor basic media could remove the chromium.
The use of saturated EDTA-Na$_2$ solution only partially extracted the chromium species. Also, ion exchange resins seemed to have an effect, although
they were not capable of fully removing the heavy metal residues. Further
attempts to precipitate chlorolactone oxime **6a** with water and isolate the
chromium free compound by filtration failed, as the impurities precipitated
as well. Recrystallisation also proved to be an inappropriate method, as
no solvent system was found, that selectively dissolved either chlorolactone
oxime **6a** or the chromium impurities. Furthermore, attempts to bind the
chromium already during the reaction were carried out. These included
addition of EDTA or ion exchange resins to the reaction mixture before hydroxylamine hydrochloride was added. However, none of them was sufficiently

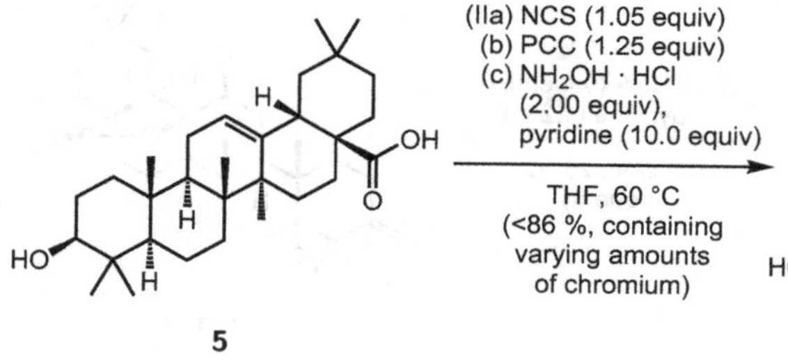

Scheme 2.10: One-pot synthesis of oxime **6a**. Reagents and conditions: (II) THF, 60 °C. (a) NCS, 1 h. (b) PCC, 3 h. (c) Hydroxylamine hydrochloride, pyridine, 5 h.

binding the chromium. As this issue could not be resolved satisfyingly, the search for alternatives to PCC became necessary.

A recently reported method by Dip et al. provided a solution to this problem. Their work demonstrates the selective oxidation of secondary alcohols with trichloroisocyanuric acid (TCCA) (Scheme 2.11), a chemical widely used for swimming pool disinfection.[4] In the general procedure a solution of TCCA (0.4 equiv, equals 1.2 equiv Cl$^+$) in EtOAc was added to a solution of the respective alcohol (1 equiv) in EtOAc at room temperature. Pyridine (1.2 equiv) was added to prevent α-chlorination. cyanuric acid, the only formed by-product, can be easily removed by filtration. After standard aqueous workup, the product is usually obtained without requirement for further purification.

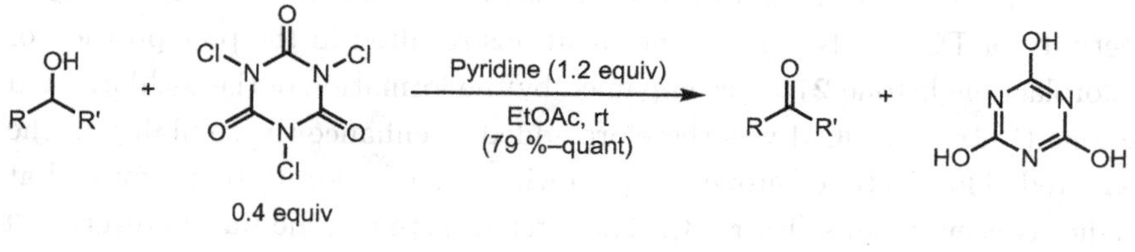

Scheme 2.11: The oxidation of secondary alcohols with TCCA, reported by Dip et al.

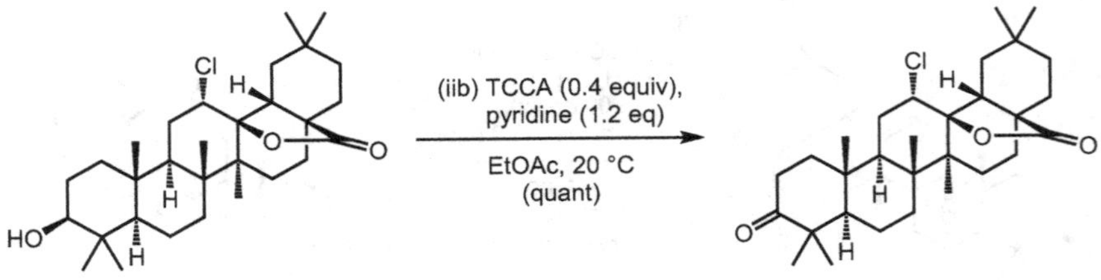

Scheme 2.12: Oxidation of chlorolactone alcohol **26a** with TCCA. Reagents and conditions: (iib) TCCA, pyridine, EtOAc, 20 °C, 3 h.

A control experiment was carried out, treating chlorolactone alcohol **26a** in EtOAc with a solution of TCCA in EtOAc. The experiment showed, that this procedure was also suitable for oxidation of the C-3 alcohol in the oleanane scaffold (Scheme 2.12).

The use of TCCA would further simplify the reaction setup, as it is not only capable of performing the oxidation, but also enabling chlorolactonisation.[42] In principle, 0.8 equiv of TCCA are necessary to achieve this tandem transformation.

Considering a potential consecutive concatenation of the oximation step, the utilisation of TCCA required several adaptations of the reaction conditions. The high reactivity of TCCA towards THF on the one hand,[43] and the poor solubility of the later targeted oxime **6a** in neat EtOAc on the other hand, demanded further solvent screening (Table 2.1). Similar conditions as those reported by Dip et al. were thereby used as a reference (Entry 1). It was found that concentration plays a crucial role in the chemoselectivity of TCCA. Too high concentrations resulted in the precipitation of chlorolactone ketone **27a**, accompanied by the formation of the α-chlorinated ketone (Entry 2). DCM was therefore added to enhance the solubility of the targeted chlorolactone ketone **27a**, allowing the reaction to be performed at higher concentrations (Entry 3). The later targeted oxime **6a**, however, still remained insoluble in the used mixture of DCM/EtOAc (1:3). Due to the previously mentioned poor solubility of oleanolic acid (**5**) in neat DCM, it was at first not considered as solvent for this reaction, although it sufficiently

Table 2.1: Solvent screening for the tandem lactonisation-oxidation with
TCCA. All reactions were carried out on a 1 mmol scale (1.0 equiv
of oleanolic acid (**5**), 0.8 equiv of TCCA and 2.4 equiv of pyridine)
at 20 °C for 2 h.

Entry	Solvent	V /ml	V(TCCA) /ml[a]	Note
1	EtOAc	16	2	Quantitative yield
2	EtOAc	8	2	5 % α-chlorination
3	DCM/EtOAc[b]	8	4	Quantitative yield
4	DCM	12	neat	Quantitative yield

[a] Solution of TCCA in solvent.
[b] 1:3 mixture.

solubilised the later targeted oxime **6a**. Conveniently, it was found later that
addition of pyridine promotes solubilisation of oleanolic acid (**5**) in DCM.
Furthermore, it was found that TCCA can be added neat to the reaction
mixture when using DCM as solvent, thus, enabling access to chlorolactone
ketone **27a** with minimal preparative effort (Entry 4).

With these conditions in hand, concatenation of the oximation step was
targeted. To prevent formation of highly explosive nitrogen trichloride or
chlorination of the oxime moiety, excess TCCA had to be quenched before the
addition of hydroxylamine hydrochloride.[44] The secondary alcohol *i*-PrOH
was employed as TCCA-scavenger. The optimised procedure consumes
15.2 equiv pyridine, 0.8 equiv TCCA, 0.8 equiv *i*-PrOH and 3.2 equiv hydroxylamine hydrochloride (Scheme 2.13). The whole sequence is carried
out at room temperature, requires an overall reaction time of 21 h and
affords chlorolactone oxime **6a** in 98 % yield. Nevertheless, the problems

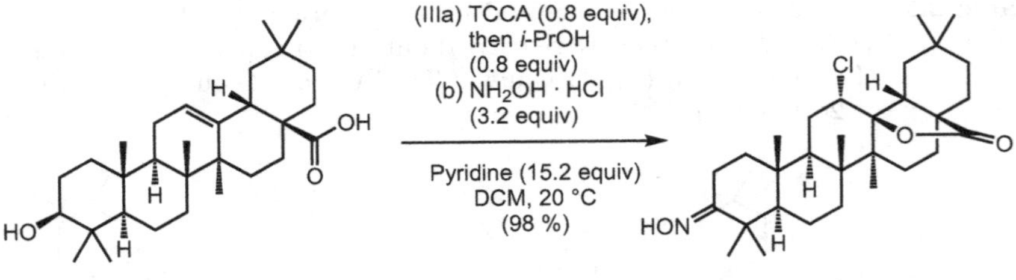

Scheme 2.13: Enhanced one-pot synthesis of chlorolactone oxime **6a**. Reagents and conditions: (III) DCM, pyridine, 20 °C. (a) TCCA, 90 min, then i-PrOH, 30 min. (b) Hydroxylamine hydrochloride, 19 h.

described later in Section 2.3.4 required an inevitable switch to the respective bromolactone.

2.3.1.2 Bromolactone Oxime

Kaminskyy et al. have previously reported a 3-step sequence for the synthesis of bromolactone oxime **6b** starting from oleanolic acid (**5**) (Scheme 2.14).[5] First, lactonisation was achieved by treatment of oleanolic acid (**5**) with bromine.[36] The oxidation was then accomplished *via* reaction of bromolactone alcohol **26b** with chromium trioxide and pyridine. Unfortunately, neither of

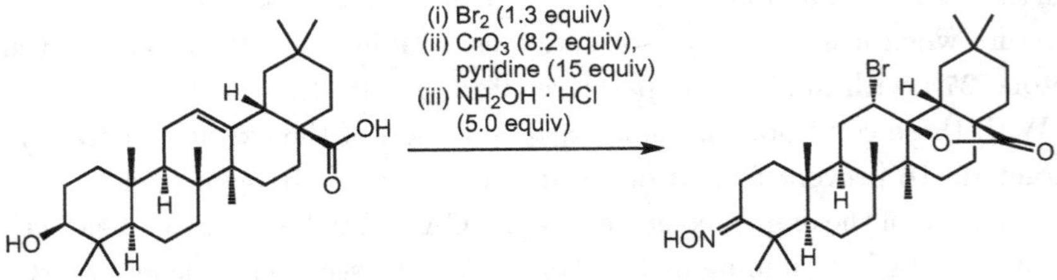

Scheme 2.14: Synthesis of bromolactone oxime **6b** reported by Kaminskyy et al.

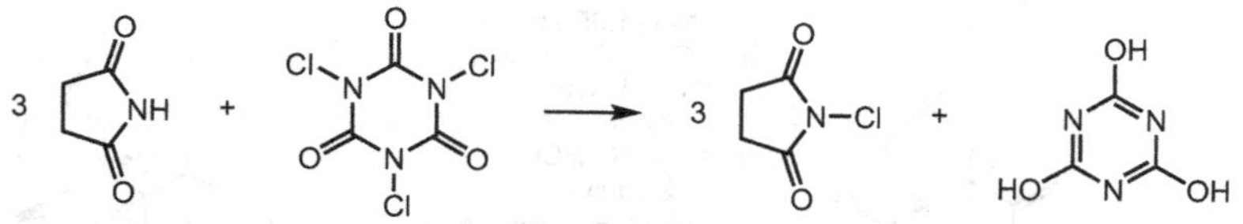

Scheme 2.15: Proposed chlorination of succinimide.

these steps has reported yields. Treatment with hydroxylamine hydrochloride gave the oxime **6b** in 91 % yield from bromolactone ketone **27b**.

Alternative conditions for lactonisation and oxidation have been reported by Elsayed et al. and Rivero-Chan et al., respectively. The former used *N*-bromosuccinimide (NBS) to form bromolactone alcohol **26b** in a yield of 76 %.[45] The latter synthesised bromolactone ketone **27b** *via* Jones oxidation.[41]

The herein developed chlorolactone oxime **6a** synthesis, described in Section 2.3.1.1, was adapted to synthesise bromolactone oxime **6b**. Initial considerations suggested the substitution of 0.4 equiv of TCCA with 1.05 equiv of NBS. Interestingly, the first attempts to synthesise bromolactone oxime **6b** according to the adapted sequence failed. Only traces of ketone were formed after addition of TCCA. Comparison of the redox potentials of NCS and TCCA may explain this observation. Due to its higher oxidation potential,[42] TCCA reoxidises the NBS waste-product, succinimide, to NCS (Scheme 2.15).

A control experiment was carried out to prove this proposal. Treatment of bromolactone alcohol **26b** with NCS did not yield bromolactone ketone **27b**

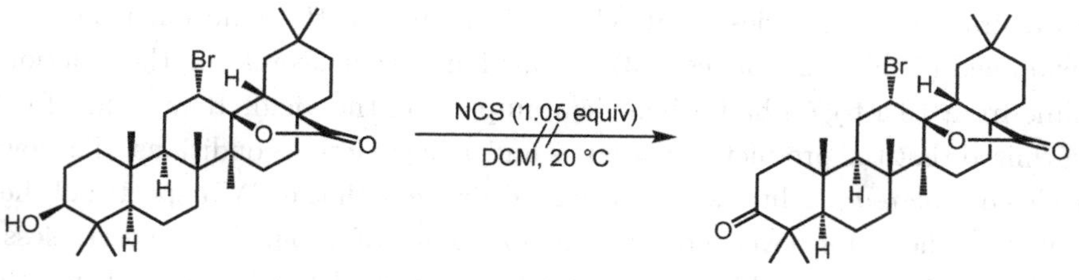

Scheme 2.16: Oxidation attempt with NCS.

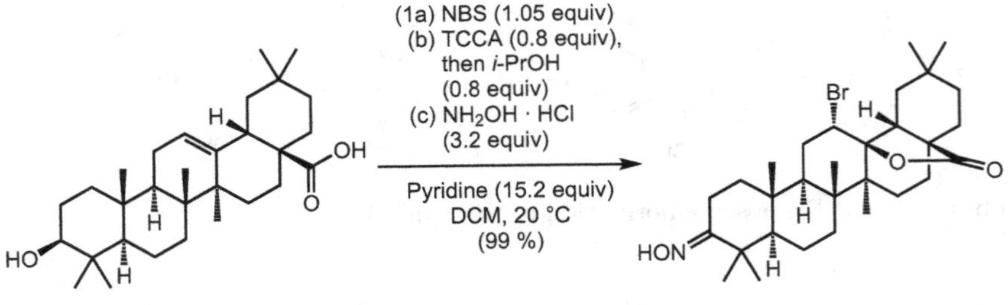

Scheme 2.17: One-pot synthesis of bromolactone oxime **6b**. Reagents and conditions: See Scheme 2.25 on Page 37.

(Scheme 2.16). Hence, 0.35 equiv of TCCA (Equals 1.05 equiv of Cl^+) are most likely quenched by the formed succinimide, preventing the oxidation of the C-3 hydroxy group. To resolve this issue, the amount of TCCA was increased to 0.8 equiv.

The optimised procedure consumes 15.2 equiv pyridine, 1.05 equiv NBS, 0.8 equiv TCCA, 0.8 equiv i-PrOH and 3.2 equiv hydroxylamine hydrochloride (Scheme 2.17). The whole sequence is carried out at room temperature, requires an overall reaction time of 24 h and affords bromolactone oxime **6b** in 99 % yield.

2.3.2 C–H Activation at position C-23

2.3.2.1 Palladium Catalysed C–H Acetoxylation

In a first reaction series at 80 °C, the influence of the reaction time was examined (Table 2.2, entries 1–3). It was found that decreasing the reaction time from 16 h to 7 h had a beneficial impact on the yield. It was therefore surmised that the product decomposes under the reaction conditions. The low yields of isolated product were not in accordance with the NMR spectra of the crude, as the latter never indicated high amounts of impurities. Nevertheless, each reaction formed high amounts of unidentified black material, which was removed by column chromatography. It was considered that this black

Table 2.2: Screening of the C–H acetoxylation of chlorolactone oxime **6a**. All reactions were carried out on a 0.1 mmol scale (1.0 equiv of chlorolactone oxime **6a**, 26.5 equiv of Ac_2O) in 0.25 ml of AcOH.

Entry	Pd(OAc)$_2$ /%	PhI(OAc)$_2$ /equiv	T /°C	t /h	X /%[a]	Y /%	dr[a]
1	10	1.5	80	16	—	31	70:30
2	10	1.5	80	15	—	29	70:30
3	10	1.5	80	7	—	34	70:30
4	20	1.5	35	29	86	47[a]	79:21
6	20	1.5	25	29	86	43[a]	77:23
5	20	2.0	25	29	83	42[a]	77:23
7	12.5	1.5	20	48	—	34	80:20

[a] Determined *via* ^{1}H-NMR spectroscopy.

material might be decomposed product, prompting an investigation of the effects of lower temperatures on the yield.

At that point, Berger successfully employed the method at 45 °C.[33] At this temperature, no black material was observed in the reaction mixture. The reaction was thereafter performed at even lower temperatures, however with higher catalyst loadings (Table 2.2, entries 4–7). The general trend observed was that higher *dr* values were obtained at lower temperatures. The highest yield was achieved at 35 °C after 29 h reaction time (Entry 4). No difference in reactivity was observed between the usage of 1.5 equiv and 2.0 equiv of PhI(OAc)$_2$ (Entries 5 and 6). To test the limits of this reaction another batch was conducted at 20 °C (Entry 7). Intriguingly, the substrate

Table 2.3: Solvent screening for C–H acetoxylation of chlorolactone O-acetyl-
oxime **28a**. All reactions were carried out on a 0.1 mmol scale
(1.0 equiv of chlorolactone O-acetyloxime **28a**) in 0.5 ml of AcOH at
80 °C.

Entry	Solvent	Pd(OAc)$_2$ /%	PhI(OAc)$_2$ /equiv	Ac$_2$O /equiv	t /h	Y /%	dr^a
1	DCE	10	1.5	10	38	<5^a	85:15
2	DCE	20	3.0	10	38	19	90:10
3	MeNO$_2$	10	1.5	10	15	<5^a	80:20
4	DCE	10	1.5	0	16	<5^a	85:15
5	MeNO$_2$	10	1.5	0	16	<5^a	80:20

^{a} Determined via ^{1}H-NMR spectroscopy.

was successfully acetoxylated, but at very slow reaction rates, as expected.
Nevertheless, the yield of 34 %, previously achieved at 80 °C was also reached
at 20 °C within 48 h (Entries 3 and 7).

Further investigations were carried out on pre-acetylated chlorolactone
oxime **28a** (Tables 2.3 and 2.4), prepared by Berger.[33] This allowed the
omission of Ac$_2$O from the reaction mixture, and therefore enabled the
screening of alternative solvent compositions.

MeNO$_2$ and DCE were examined as alternative solvents. The reactions
were set up with and without Ac$_2$O additive respectively (Table 2.3). Notably,
exchanging AcOH for either MeNO$_2$ or DCE drastically reduced the desired
reactivity. Only doubling of the catalyst loading and the amount of oxidant
gave a reasonable yield of 19 %, alongside an improved dr of 90:10 (Entry
2).

Table 2.4: Screening of the C–H acetoxylation of chlorolactone *O*-acetyloxime
28a. All reactions were carried out on a 0.1 mmol scale (1.0 equiv of
chlorolactone *O*-acetyloxime **28a**, 0.20 equiv of $Pd(OAc)_2$, 1.2 equiv
of $Mn(OAc)_2$ and 5.0 equiv of Oxone) in 0.25 ml of solvent at 80 °C.

Entry	Solvent	Ac_2O/equiv	t/h	Y/%	dr^a
1	DCE	10	16	<5	80:20
2	DCE	0	16	0	—
3	$MeNO_2$	10	40	<5	—
4	$MeNO_2$	0	40	<5	—

a Determined *via* ^{1}H-NMR spectroscopy.

Reddy et al. previously reported a palladium catalysed acetoxylation with
$Mn(OAc)_2$ as co-catalyst and Oxone as oxidant.[30] Several attempts to apply
this method have been carried out (Table 2.4). However, only traces of the
desired acetoxylated product **7a** were observed (Entries 1, 3 and 4). When
the reaction was performed in DCE without Ac_2O, the starting material
was recovered (Entry 2).

Further examinations were carried out on *O*-methylated chlorolactone
oxime **29a**. C–H Acetoxylation of *O*-methyl oximes has been previously
reported by Sanford and coworkers.[2] The examined chlorolactone *O*-methyl
oxime **29a** was synthesised from chlorolactone ketone **27a** *via* oximation
with methoxyamine hydrochloride.

The C–H acetoxylation of chlorolactone *O*-methyl oxime **29a** was examined
in AcOH and DCE (Table 2.5). Neither of the two methods yielded the
acetoxylated chlorolactone *O*-methyl oxime **30a**.

Table 2.5: Screening of the C–H acetoxylation of chlorolactone *O*-methyloxime **29a**. All reactions were carried out on a 0.1 mmol scale (1.0 equiv of *O*-methyloxime **29a**, 0.2 equiv of $Pd(OAc)_2$ and 1.2 equiv of $PhI(OAc)_2$) at 75 °C for 8 h.

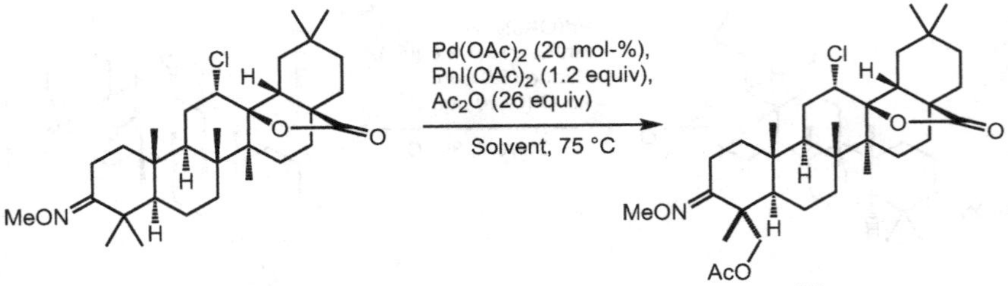

29a **30a**

Entry	Solvent	V(Solvent) /ml	Ac_2O /equiv	Note
1	AcOH	0.25	26.5	No target molecule
2	DCE	0.50	10.0	No target molecule

Due to the issues described later in Section 2.3.4, it was thereafter switched to the respective bromolactone **6b**. Berger slightly modified the C–H acetoxylation conditions described in Table 2.3.[33] C–H Acetoxylation of bro-

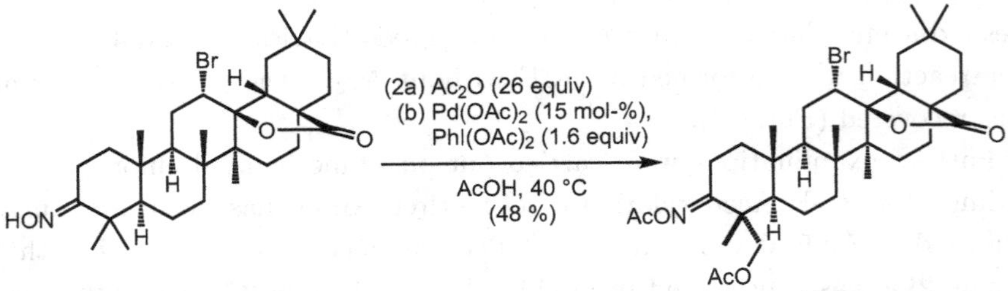

6b **7b**
(*dr* 75:25)

Scheme 2.18: Synthesis of **7b**. Reagents and conditions: See Scheme 2.25 on Page 37.

Table 2.6: Screening of the C–H acetoxylation of bromolactone oxime **6b**. All reactions were carried out with 0.15 equiv of Pd(OAc)$_2$ and 26 equiv of Ac$_2$O in AcOH (2.5 ml mmol^{-1}) for 16 h.

Entry	$n($**6b**$)$/mmol	PhI(OAc)$_2$/equiv	T/°C	Y/%	dr[a]
1	20	1.6	40	48	75:25
2	1	1.8	40	40	76:24
3	1	1.8	50	38	75:25

[a] Determined *via* ^{1}H-NMR spectroscopy.

molactone oxime **6b** was carried out according to the conditions optimised by Berger (Scheme 2.18).

At last, the correlation between reaction temperature and yield was studied (Table 2.6). Increasing the temperature from 40 °C to 50 °C already resulted in a negative impact on the yield (Entries 2 and 3). Notably, a 20 mmol scale reaction gave a higher yield than a 1 mmol scale reaction (Entries 1 and 2).

2.3.2.2 Copper Catalysed C–H Hydroxylation

The Schönecker oxidation was examined as alternative method for installation of the C-23 hydroxy group. Successful implementation would shorten the overall synthetic sequence by one step, as no separate directing group cleavage would be needed to generate β-hydroxy ketone **8a**. To apply this method, chlorolactone ketone **27a** had to be transformed into the respective picolylimine **31a** first. Several attempts to employ the procedure recently reported by Baran and coworkers,[46] on chlorolactone ketone **27a** failed.

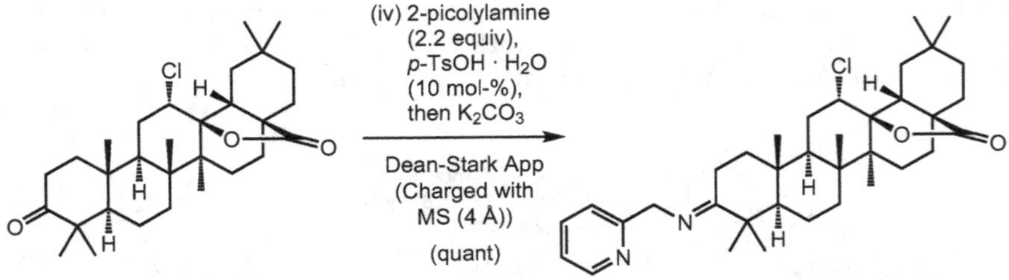

Scheme 2.19: Synthesis of picolylimine **31a**. Reagents and conditions: (iv)
Dean-Stark apparatus (charged with activated molecular sieves
(4 Å)), 2-picolylamine, p–TsOH · H_2O, toluene, reflux, 8 h, then
20 °C, K_2CO_3.

Initial assumptions that the neighbouring *gem*-dimethyl moiety would
sterically hinder the formation of the imine were not confirmed. In fact, a
rather fast hydrolysis was the reason why isolation of picolylimine **31a** was
unsuccessful. Picolylimine **31a** even hydrolysed rapidly in chloroform, after
it was filtered over basic aluminium oxide and stored over molecular sieves.
Measuring NMR samples over solid K_2CO_3 was found to be an appropriate
method to slow down hydrolysis.

Adaptation of the procedure described by Baran and coworkers required
loading of the Dean-Stark apparatus with activated molecular sieves, and
furthermore omission of the aqueous workup (Scheme 2.19). K_2CO_3 was
added to quench the reaction. Removal of p–TsOH was achieved by precipitation with dry Et_2O. Inert filtration and subsequent removal of volatiles
afforded imine **31a** in quantitative yield. Imine **31a** is stable for several
months, if stored dry under an argon atmosphere at +5 °C.

Baran and coworkers optimised the C–H hydroxylation protocol in a
mixture of MeOH and acetone.[46] Considering the rapid hydrolysis of picolylimine **31a** it is hardly surprising that the compound was not stable
under these conditions (Table 2.7, entries 1 and 2). With MeOH as a protic
solvent, and acetone as a potential amine scavenger, only hydrolysed ketone
27a was isolated. The effect of added K_2CO_3 and $CaCO_3$, in conjunction
with different solvents was examined (Entries 3–8). The desired β-hydroxy

Table 2.7: Screening of the Schönecker oxidation. All reactions were carried out on a 0.1 mmol scale (1.0 equiv of picolylimine **31a**, 1.3 equiv of [Cu(MeCN)$_4$][PF$_6$], 2.0 equiv of sodium L-ascorbate and O$_2$ atmosphere) at 50 °C.

Entry	Solvent	V /ml	Additive	n(Additive) /equiv	Note
1	MeOH/AcMe[a]	0.5	—	—	Hydrolysed
2	AcMe	0.5	—	—	Hydrolysed
3	MeOH/AcMe[a]	0.8	K$_2$CO$_3$	2.0	Hydrolysed
4	MeOH/AcMe[a]	0.8	CaCO$_3$	2.0	Hydrolysed
5	MeCN	0.8	CaCO$_3$	2.0	Hydrolysed
6	DMF	0.8	CaCO$_3$	2.0	Hydrolysed
7	DMSO	0.8	CaCO$_3$	2.0	Hydrolysed
8	THF	1.6	CaCO$_3$	2.0	Hydrolysed

[a] 1:1 mixture.

ketone **8a** was not detected with mass spectroscopy in any of these reactions. Thus, the preparation of β-hydroxy ketone **8a** *via* Schönecker oxidation was not successful.

2.3.3 Deacetylation and Deoximation

Deacetylation and deoximation were carried out in a one-pot reaction as elaborated by Berger (Section 2.2.3), giving β-hydroxy ketone **8b** as a single diastereomer in 55 % yield.

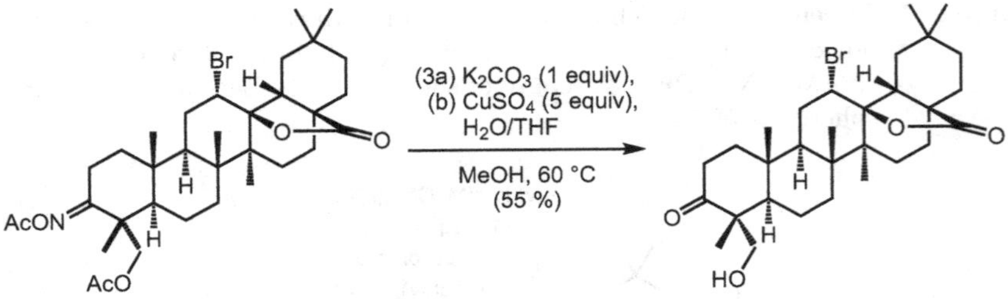

7b (*dr* 75:25) **8b** (*dr* > 20:1)

Scheme 2.20: One-pot deacetylation-deoximation of bromolactone **7b**. Reagents and conditions: See Scheme 2.25 on Page 37.

The low yield of this deprotection prompted further investigations. During examination of reaction conditions, it was found that longer reaction times caused consumption of the desired β-hydroxy ketone **8b**. The newly formed by-product was identified as dehydroxymethylated molecule **32b**, generated in a retro-aldol reaction (Scheme 2.21). This reactivity has been observed previously in 23-hydroxy-3-oxo-oleananes and 24-hydroxy-3-oxo-oleananes.[9,47,48]

Barton and Mayo described the mechanisms for this transformation under acidic as well as under basic conditions.[47] The herein reported setup features the simultaneous presence of a Lewis acid (Cu^{2+}) and a base ($CO_3{}^{2-}$). Hence, the β-hydroxy ketone can most likely be dually activated (Scheme 2.22). After

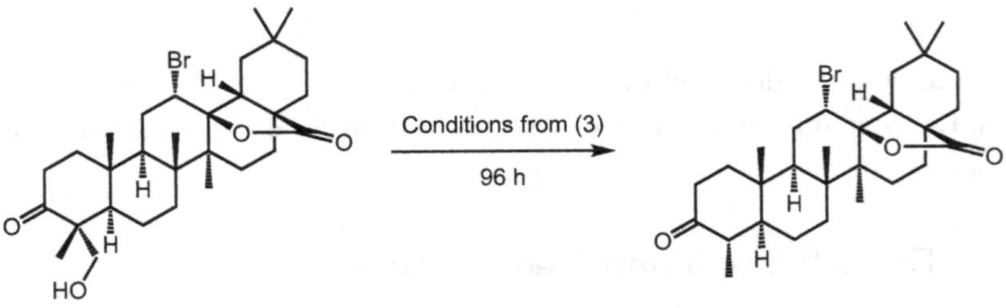

8b **32b**

Scheme 2.21: Decarbonylation of β-hydroxy ketone **8b**

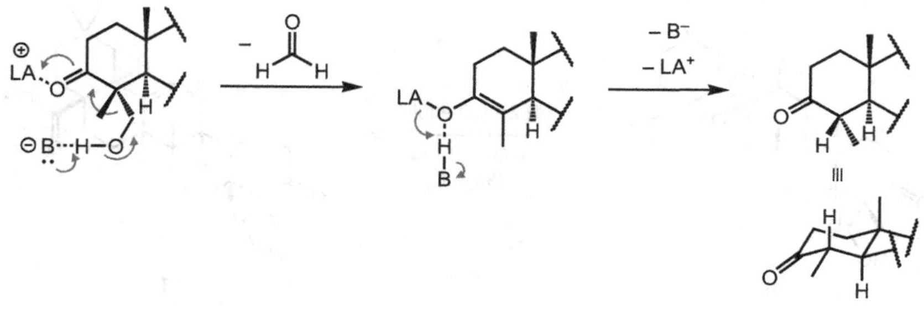

Scheme 2.22: Dually activated retro-aldol formaldehyde elimination.

elimination of one molecule of formaldehyde, the formed enolate intermediate is protonated to afford the thermodynamically favoured epimer **32b** with α-equatorial conformation.[9]

2.3.4 Retro-halolactonisation

It was not foreseen, that the initial reason for using the chlorolactone – the higher C–X bond energy – would ultimately doom the synthetic route to failure.

Following up the retro-chlorolactonisation attempts carried out by Berger (Section 2.2.4), the suitability of a retro-chlorolactonisation method reported by Soengas et al. was examined.[49] Their procedure involves magnesium, zinc chloride and catalytic iodine in THF. Applied to chlorolactone ketone

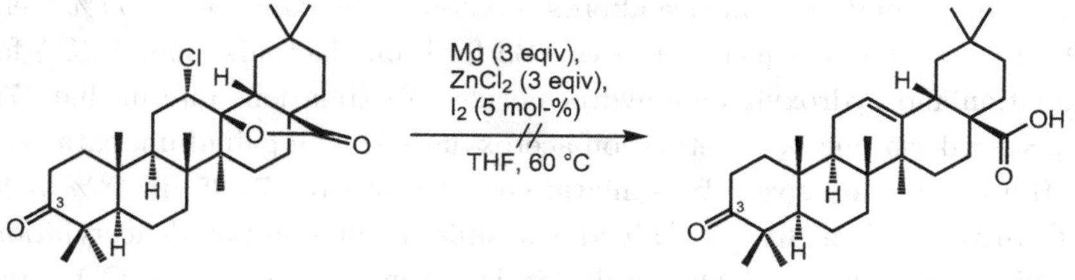

Scheme 2.23: Failed retro-chlorolactonisation of **27a**.

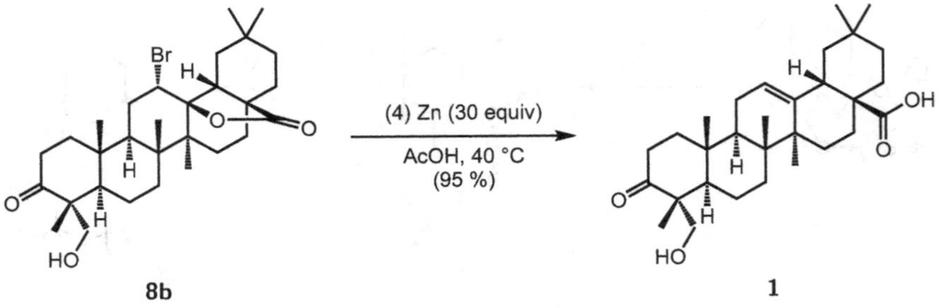

Scheme 2.24: Retro-bromolactonisation of **8b**. Reagents and conditions: See Scheme 2.25 on Page 37.

27a, however, this method was not successful. Cleavage of the chlorolactone thus failed with all examined methods (Scheme 2.23 and Section 2.2.4).

At that point, the choice of halolactone had to be reconsidered. As a result, the synthetic sequence was repeated with the respective bromolactone. Applying the procedure reported by Lewis and Tucker to bromolactone hydroxyketone **8b** gave the pure target molecule hederagonic acid (**1**) after column chromatography in 95 % yield.

2.4 Final Synthetic Route to Hederagonic Acid

The following section shortly summarises the complete semisynthesis of hederagonic acid (Scheme 2.25). The synthetic route starts from oleanolic acid (**5**). Application of the herein described one-pot bromolactonisation-oxidation-oximation sequence affords bromolactone oxime **6b** in 99 % yield. The 3-step one-pot sequence utilises NBS for bromolactonisation, TCCA for oxidation and hydroxylamine hydrochloride for oximation. Intermediate **7b** is accessed *via* one-pot acetylation-acetoxylation. The palladium catalysed C–H acetoxylation gives **7b** as mixture of 4-epimers (*dr* 75:25) in 48 % yield. β-Hydroxy ketone **8b** is obtained *via* subsequent one-pot deacetylation-deoximation sequence. Deacetylation is accomplished with K_2CO_3, and deoximation with $CuSO_4$. The undesired 4-epimer was removed *via* column chromatography, giving β-hydroxy ketone **8b** as a single diastereomer in 55 % yield. The final retro-halolactonisation is achieved with zinc in acetic acid, affording the target product **1** in 95 % yield. Hederagonic acid (**1**) is thus synthesised in 4 steps and 20 % overall yield.

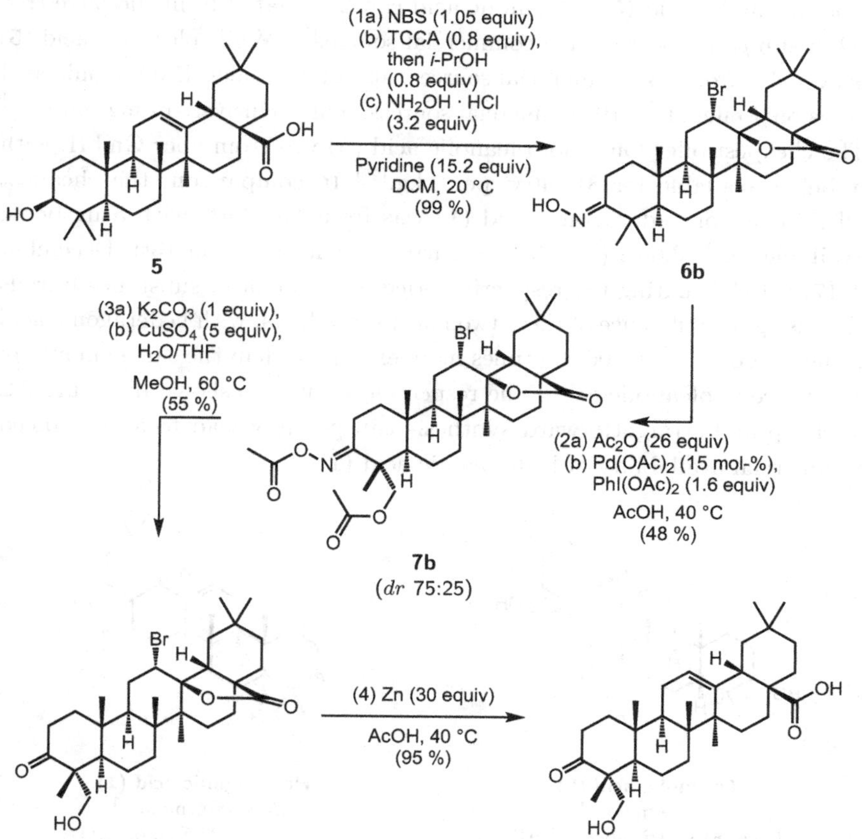

Scheme 2.25: The herein reported semisynthesis of hederagonic acid (**1**). Reagents and conditions: (1) DCM, pyridine, 20 °C. (a) NBS, 60 min. (b) TCCA, 90 min, then *i*-PrOH, 30 min. (c) Hydroxylamine hydrochloride, 21 h. (2) AcOH, 40 °C. (a) Ac₂O, 2 h. (b) PhI(OAc)₂, Pd(OAc)₂, 16 h. (3) MeOH, 60 °C. (a) K₂CO₃, 1 h. (b) THF, CuSO₄·5 H₂O in H₂O, 16 h. (4) Zn, AcOH, 40 °C, 2 h.

2.4.1 Economic Considerations

This work presented a 4-step synthetic route to hederagonic acid (**1**) starting from oleanolic acid (**5**). Oleanolic acid can be isolated from the olive tree (*Olea europaea*), a profitable commercial source.[50] While oleanolic acid (**5**) is available from 134 commercial sources, also on kg scales, hederagonic acid (**1**) is only offered by 19 commercial sources, and exclusively in mg units.[51] The cheapest offer found for oleanolic acid (**5**) was from abcr GmbH, with it being available for 816.40 € per 1 kg.[52] In comparison, the cheapest offer found for hederagonic acid (**1**) was from MolPort, with 5 mg being available for 513.00 \$ (436.77 € – Currency conversion on 13th December 2017).[53] Calculating the respective price per amount of substance unveils the vast price difference of these two natural products, with hederagonic acid being approximately 100 000 times more expensive than the same amount of substance of oleanolic acid. The respective values are shown in Figure 2.2. A cheap and straightforward synthesis can possibly lead to an enhanced commercial availability of hederagonic acid (**1**).

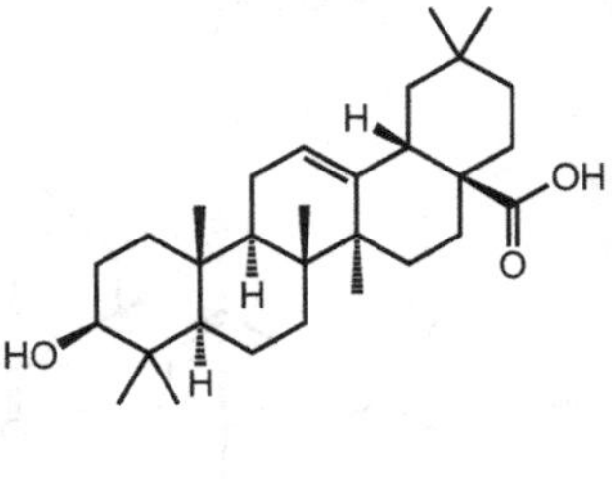

Oleanolic acid (5)
0.39 € mmol^{-1}
(1 kg, 816.40 €, abcr GmbH)

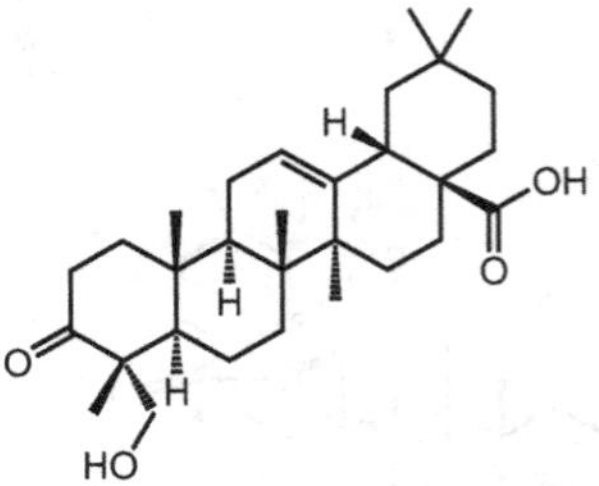

Hederagonic acid (1)
42 000.00 € mmol^{-1}
(5 mg, 436.77 €, MolPort)

Figure 2.2: Commercial prices of oleanolic acid (**5**) and hederagonic acid (**1**).

2.5 Summary and Outlook

A one-pot reaction was designed for the synthesis of the intermediate oxime. This reaction combines three single steps, and utilises TCCA as oxidant. The one-pot sequence can be carried out at room temperature and affords the intermediate oxime in 99 % yield (1 mmol batch, 97 % in a 20 mmol batch). The use of TCCA allowed omission of commonly used PCC, a known carcinogen. The obtained oxime can be used after aqueous workup, without further purification.

A palladium catalysed C–H acetoxylation at 40 °C was utilised for the introduction of the C-23 hydroxy group. Acetoxylation was even successfully carried out at 20 °C, albeit in low yield. This represents the first example of catalytic C–H acetoxylation at such low temperatures. Furthermore, scalability of this reaction was demonstrated in a 20 mmol batch. As the diastereoselectivity of the C–H acetoxylation is solely substrate controlled, utilisation of chiral ligands could possibly increase the dr of this transformation.

The deacetylation-deoximation step may be improved by neutralising excess K_2CO_3. The absence of base might disfavour the described retro-aldol reactivity. Following this strategy, Berger was able to obtain 65 % yield in this transformation.[33]

In summary, the development of a novel semisynthetic approach to hederagonic acid (**1**) was presented. Deployment of carefully designed one-pot reactions allowed the shortening of the synthetic sequence, resulting in the to date shortest route to hederagonic acid (**1**).

At last, the herein synthesised β-hydroxy ketone intermediate **8b** could serve as branching point for the syntheses of more complex oleananes (Scheme 2.26). As example, further functionalisation at C-2 can lead to arjunolic acid (**33a**) and bayogenin (**33b**).[1] Potential modifications at C-6 would allow access to higher members of the oleanane family, like uncargenin C (**33c**) and terminolic acid (**33d**). The versatile biological activities of these natural products, combined with their utterly limited commercial availability, makes them interesting targets in natural product synthesis.

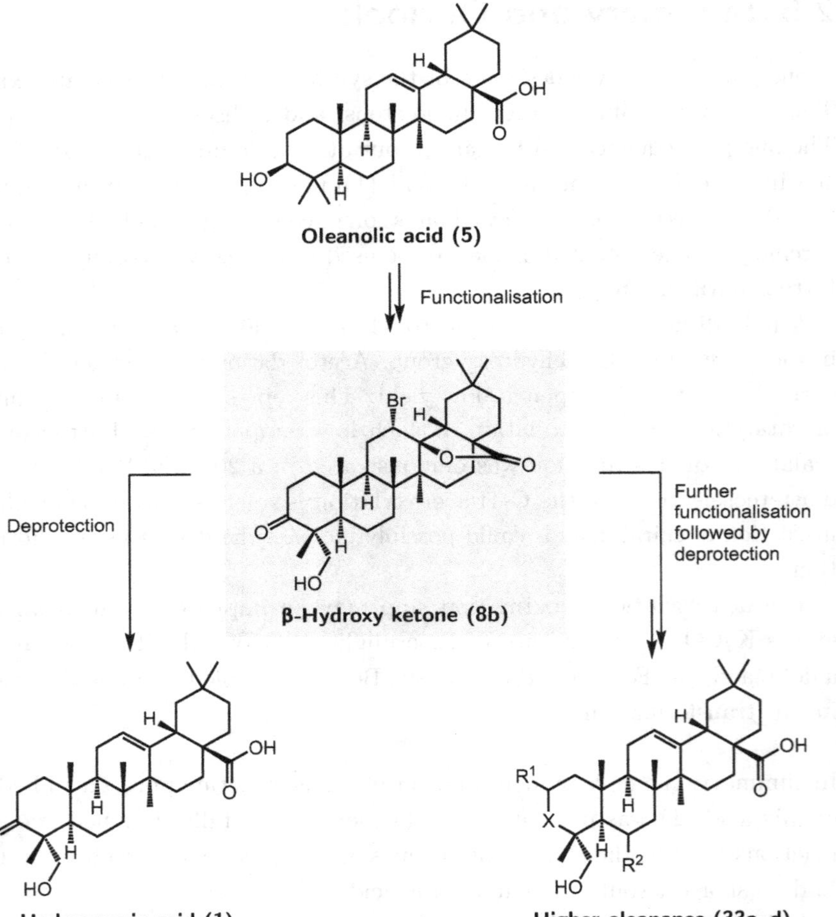

33	X	R¹	R²
a	C=O	α-OH	H
b	C=O	β-OH	H
c	CH-β-OH	H	β-OH
d	CH-β-OH	α-OH	β-OH

Scheme 2.26: β-Hydroxy ketone **8b** as potential branching point for routes to higher members of the oleanane family (**33a–d**).

3 Experimental Section

3.1 General Information

Glassware and Techniques: Unless otherwise indicated, no glassware was flame dried before use, and all reactions were performed under an air atmosphere, not using standard Schlenk-techniques. **Reagents and Solvents:** Unless otherwise stated, all reagents and solvents were used as received from commercial suppliers. **Reaction Monitoring:** Reaction progress was monitored using thin layer chromatography (TLC) on aluminium sheets coated with silica gel 60 with 0.2 mm thickness (Pre-coated TLC-sheets ALUGRAM© Xtra SIL G/UV$_{254}$). Visualisation was achieved either by UV light (254 nm and 363 nm), by treatment with potassium permanganate and heat, or by treatment with ethanolic phosphomolybdic acid and heat. **Column Chromatography:** Flash column chromatography was performed using silica gel 60 (230 to 400 Mesh, Merck and co.). **NMR Spectra:** All NMR spectra were recorded on Bruker Avance III 600 or Avance III 700 spectrometers. Chemical shifts (δ) were given in ppm, referenced to the peak of tetramethylsilane (TMS), using residual non-deuterated solvent as internal standard (^{1}H: δ(CDCl$_3$) = 7.26 ppm; ^{13}C: δ(CDCl$_3$) = 77.16 ppm).[54] Coupling constants (J) were quoted in Hz. Spectroscopy splitting patterns were designated as singlet (s), doublet (d), triplet (t), quartet (q), quintet (quint), or combinations thereof. Splitting patterns that could not be interpreted or easily visualised were designated as multiplet (m). **Mass Spectra:** Mass spectra were obtained using a Bruker maXis spectrometer with ESI-TOF. **IR Spectra:** Neat infra-red spectra were recorded using a Perkin-Elmer Spectrum 100 FT-IR spectrometer. Wavenumbers ($\tilde{\nu}$) were reported in cm^{-1}. **Optical rotation:** Specific rotation ($[\alpha]_D^{20}$) was determined on a SCHMIDT+HAENSCH UniPol L 2000 polarimeter at

© The Editor(s) (if applicable) and The Author(s), under exclusive license to
Springer Fachmedien Wiesbaden GmbH, part of Springer Nature 2020
C. Knittl-Frank, *A Concise Semisynthesis of Hederagonic Acid*, BestMasters,
https://doi.org/10.1007/978-3-658-30011-1_3

589.44 nm (sodium D line) in a cell with 100 mm path length. $[\alpha]_D^{20}$ values were reported in $10^{-1}\,\mathrm{deg\,cm^2\,g^{-1}}$. **X-Ray Analysis:** The X-ray intensity data were measured on a Bruker D8 Venture diffractometer equipped with multilayer monochromator, Cu-K$_\alpha$ INCOATEC micro focus sealed tube and Oxford cooling device. The structure was solved by direct methods and refined by full-matrix least-squares techniques. Non-hydrogen atoms were refined with anisotropic displacement parameters. Hydrogen atoms were inserted at calculated positions and refined with a riding model. The following software was used: Bruker SAINT software package using a narrow-frame algorithm for frame integration,[55] SADABS for absorption correction,[56] OLEX2 for structure solution, refinement, molecular diagrams and graphical user-interface,[57] ShelXle for refinement and graphical user-interface,[58] SHELXS-2013 for structure solution,[59] SHELXL-2013 for refinement,[58] Platon for symmetry check,[60] mercury for image compilation.[61]

3.2 Experimental Procedures

3.2.1 12α-Chloro-3β-hydroxyolean-28,13β-olide (26a)

A round bottom flask, equipped with magnetic stirring bar, was charged with a solution of oleanolic acid (**5**) (10.0 mmol, 4660 mg, 1 equiv) in dry THF (40 ml). After addition of NCS (10.5 mmol, 1402 mg, 1.05 equiv), the flask was immersed into a preheated oilbath (60 °C) and the reaction was left stirring at 60 °C for 60 min. After cooling to room temperature, the solution was diluted with EtOAc (500 ml) and washed with $Na_2S_2O_3$ solution (0.1 M in H_2O) and H_2O (9:1) (250 ml), followed by $NaHCO_3$ solution (saturated in H_2O) (250 ml). After drying over $MgSO_4$ the solution was concentrated under reduced pressure. The residue was dried *in vacuo* to give the target compound as pale yellow solid (quantitative yield). The obtained product was used without further purification.

^{1}H-NMR (600 MHz, CDCl$_3$): δ 4.20–4.16 (m, 1 H, H^{12}), 3.25 (dd, $J =$ 11.6, 4.5 Hz, 1 H, H^3), 2.25 (ddd, $J = 13.9, 12.1, 3.2$ Hz, 1 H, H^{11a}), 2.20–2.11 (m, 2 H, H^{16a,19a}), 2.05–1.98 (m, 2 H, H^{5,19a}), 1.94 (td, $J = 13.7, 6.1$ Hz, 1 H, H^{15a}), 1.78–1.51 (m, 9 H, H^{1a,2ab,6a,7a,11b,18,22ab}), 1.47–1.41 (m, 1 H, H^{6b}), 1.39 (s, 3 H, H^{27}), 1.38–1.32 (m, 1 H, H^{21a}), 1.32–1.24 (m, 3 H, H^{7b,16b,21b}), 1.24–1.17 (m, 4 H, H^{15b,26}), 1.04 (td, $J = 13.2, 3.6$ Hz, 1 H, H^{1b}), 0.99 (s, 6 H, H^{23,29}), 0.90 (s, 3 H, H^{30}), 0.88 (s, 3 H, H^{25}), 0.80–0.76 (m, 4 H, H^{5,24}). **^{13}C-NMR (150 MHz, CDCl$_3$):** δ 179.3 (C^{28}), 91.8 (C^{13}), 78.9 (C^3), 65.2 (C^{12}), 55.3 (C^5), 52.1 (C^{18}), 45.4 (C^{17}), 44.9 (C^9), 43.1 (C^{14}), 42.5 (C^8), 39.9 (C^{19}), 39.0 (C^4), 38.6 (C^1), 36.6 (C^{10}), 34.6 (C^7), 34.0 (C^{21}), 33.4 (C^{29}), 32.0 (C^{20}), 29.7 (C^{11}), 29.1 (C^{15}), 28.1 (C^{23}), 27.6 (C^{22}), 27.3 (C^2), 23.8 (C^{30}), 21.4 (C^{16}), 20.3 (C^{27}), 19.0 (C^{26}), 17.8 (C^6), 16.9 (C^{25}), 15.5 (C^{24}). **[α]$_D^{20}$** +56.5 (*c* 1.0, CDCl$_3$). **IR (neat)** 2932, 2867, 1768, 1713, 1467. **HRMS (ESI-TOF)** m/z: [M + Na]$^+$ Calcd for $C_{30}H_{47}ClNaO_3^+$ 513.3106; Found 513.3106.

3.2.2 12α-Bromo-3β-hydroxyolean-28,13β-olide (26b)

A round bottom flask, equipped with magnetic stirring bar, was charged with a solution of oleanolic acid (**5**) (1.00 mmol, 457 mg, 1.00 equiv) in DCM (12 ml) and pyridine (15.2 mmol, 1229 µl, 15.2 equiv). After addition of NBS (1.05 mmol, 187 mg, 1.05 equiv), the reaction was left stirring at 20 °C for

60 min. Then *i*-PrOH (1.60 mmol, 122 µl, 1.60 equiv) was added, and the mixture was left stirring at 20 °C for 30 min. The solution was subsequently diluted with DCM (20 ml) and washed with HCl solution (1 M in H_2O) (32 ml). The aqueous phase was extracted with DCM (2 × 16 ml). The combined organic layers were washed with NaOH solution (1 M in H_2O) (32 ml), followed by NH_4Cl solution (half-saturated in H_2O) (16 ml). After drying over $MgSO_4$ the solution was concentrated under reduced pressure. The residue was dried *in vacuo* to give the target compound as pale yellow solid (quantitative yield). The obtained product was used without further purification. The spectroscopic data is in accordance with the literature.[36,62]

^{1}H-NMR (600 MHz, CDCl$_3$): δ 4.30 (dd, $J = 3.6, 2.4$ Hz, 1 H, H^{12}), 3.26 (dd, $J = 11.6, 4.4$ Hz, 1 H, H^3), 2.40–2.31 (m, 2 H, H^{11a,19a}), 2.16 (td, $J = 13.4, 5.7$ Hz, 1 H, H^{16a}), 2.04–1.92 (m, 3 H, H^{15a,18,19a}), 1.85 (dt, $J = 15.0, 2.3$ Hz, 1 H, H^{11b}), 1.77–1.73 (m, 1 H, H^9), 1.72–1.60 (m, 5 H, H^{1a,2ab,22ab}), 1.60–1.51 (m, 2 H, H^{6a,7a}), 1.47–1.40 (m, 4 H, H^{6b,27}), 1.40–1.24 (m, 4 H, H^{7b,16a,21ab}), 1.23–1.17 (m, 4 H, H^{15b,26}), 1.11–1.04 (m, 1 H, H^{1b}), 1.00 (s, 6 H, H^{23,29}), 0.90 (s, 3 H, H^{30}), 0.88 (s, 3 H, H^{25}), 0.79 (dd, $J = 12.1, 1.7$ Hz, 1 H, H^5), 0.77 (s, 3 H, H^{24}). **^{13}C-NMR (150 MHz, CDCl$_3$):** δ 179.1 (C^{28}), 91.8 (C^{13}), 78.8 (C^3), 56.6 (C^{12}), 55.3 (C^5), 52.5 (C^{18}), 45.7 (C^9), 45.5 (C^{17}), 43.5 (C^{14}), 42.5 (C^8), 40.1 (C^{19}), 39.0 (C^4), 38.4 (C^1), 36.7 (C^{10}), 34.7 (C^7), 34.0 (C^{21}), 33.4 (C^{29}), 32.0 (C^{20}), 30.6 (C^{11}), 29.3 (C^{15}), 28.1 (C^{23}), 27.6 (C^{22}), 27.3 (C^2), 23.7 (C^{30}), 21.4 (C^{16}), 21.3 (C^{27}), 19.2 (C^{26}), 17.8 (C^6), 17.1 (C^{25}), 15.5 (C^{24}). $[\alpha]_D^{20}$ +66.6 (*c* 1.0, CDCl$_3$). **HRMS (ESI-TOF)** m/z: [M + Na]$^+$ Calcd for $C_{30}H_{47}BrNaO_3{}^+$ 557.2601; Found 557.2591.

3.2.3 12α-Chloro-3-oxo-olean-28,13β-olide (27a)

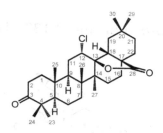

A round bottom flask, equipped with magnetic stirring bar, was charged with a suspension of oleanolic acid (5) (1.00 mmol, 457 mg, 1.00 equiv) in EtOAc (16 ml) and pyridine (2.40 mmol, 194 µl, 2.40 equiv). After addition of a solution of trichloroisocyanuric acid (0.80 mmol, 186 mg, 0.80 equiv) in EtOAc (2 ml) the mixture was left stirring at 20 °C for 2 h. The solution was filtered over a glass frit (P 16) and the residue was washed with EtOAc (2 ml). The organics were washed with HCl solution (1 M in H_2O) (5 ml), $NaHCO_3$ solution (saturated in H_2O) (5 ml) and NaCl solution (saturated in H_2O) (5 ml). After drying over $MgSO_4$ the solution was concentrated under reduced pressure. The residue was dried *in vacuo* to give the target compound as colourless foam (quantitative yield).

^{1}H-NMR (600 MHz, CDCl$_3$): δ 4.19 (dd, $J = 3.8, 2.4$ Hz, 1 H, H^{12}), 2.50 (dd, $J = 8.7, 6.4$ Hz, 2 H, H^2), 2.30 (ddd, $J = 14.7, 12.6, 3.8$ Hz, 1 H, H^{11a}), 2.20–2.13 (m, 2 H, H^{16a,19a}), 2.06–1.99 (m, 2 H, H^{18,19b}), 1.99–1.89 (m, 2 H, H^{1a,15a}), 1.85 (dd, $J = 12.6, 2.0$ Hz, 1 H, H^9), 1.74 (dt, $J = 14.8, 2.2$ Hz, 1 H, H^{11b}), 1.65 (dd, $J = 9.0, 3.5$ Hz, 2 H, H^{22ab}), 1.62–1.44 (m, 5 H, H^{1b,5,6ab,7a}), 1.40 (s, 3 H, H^{27}), 1.38–1.32 (m, 2 H, H^{21ab}), 1.32–1.26 (m, 2 H, H^{7b,16b}), 1.24 (s, 4 H, H^{15b,26}), 1.10 (s, 3 H, H^{23}), 1.04 (s, 3 H, H^{24}), 1.00 (s, 3 H, H^{29}), 0.97 (s, 3 H, H^{25}), 0.91 (s, 3 H, H^{30}). **^{13}C-NMR (150 MHz, CDCl$_3$):** δ 217.8 (C^3), 179.1 (C^{28}), 91.7 (C^{13}), 65.0 (C^{12}), 54.6 (C^5), 52.1 (C^{18}), 47.3 (C^4), 45.4 (C^{17}), 44.2 (C^9), 43.2 (C^{14}), 42.3 (C^8), 39.9 (C^{19}), 39.4 (C^1), 36.3 (C^{10}), 34.0 (C^{21}), 34.0 (C^2), 33.9 (C^7), 33.4 (C^{29}), 32.0 (C^{20}), 29.9 (C^{11}), 29.1 (C^{15}), 27.6 (C^{22}), 27.1 (C^{23}), 23.7 (C^{30}), 21.4 (C^{16}), 21.1 (C^{24}), 20.2 (C^{27}), 19.2 (C^6), 18.6 (C^{26}), 17.0 (C^{25}). $[\alpha]_D^{20}$ +76.4 (c 1.0, CDCl$_3$). **IR**

(neat) 2953, 2867, 1776, 1704, 1463. **HRMS (ESI-TOF)** m/z: $[M + Na]^+$
Calcd for $C_{30}H_{45}ClNaO_3^+$ 511.2949; Found 511.2949.

3.2.4 12α-Bromo-3-oxo-olean-28,13β-olide (27b)

A round bottom flask, equipped with magnetic stirring bar, was charged
with a solution of oleanolic acid (**5**) (1.00 mmol, 457 mg, 1.00 equiv) in DCM
(12 ml) and pyridine (15.2 mmol, 1229 μl, 15.2 equiv). After addition of
NBS (1.05 mmol, 187 mg, 1.05 equiv), the reaction was left stirring at 20 °C
for 60 min. Subsequently, trichloroisocyanuric acid (0.80 mmol, 186 mg,
0.80 equiv) was added, and the mixture was left stirring at 20 °C for 90 min.
Then *i*-PrOH (1.60 mmol, 122 μl, 1.60 equiv) was added, and the mixture
was left stirring at 20 °C for 30 min. The solution was diluted with DCM
(12 ml), filtered over a glass frit (P 16) and the residue was washed with
DCM (2 × 4 ml). The organics were washed with HCl solution (1 M in H_2O)
(32 ml). The aqueous phase was extracted with DCM (2 × 16 ml). The
combined organic layers were washed with NaOH solution (1 M in H_2O)
(32 ml), followed by NH_4Cl solution (half-saturated in H_2O) (16 ml). After
drying over $MgSO_4$ the solution was concentrated under reduced pressure.
The residue was dried *in vacuo* to give the target compound as pale yellow
foam (quantitative yield). The obtained product was used without further
purification. The spectroscopic data is in accordance with the literature.[36,41]
¹H-NMR (600 MHz, CDCl₃): δ 4.30 (t, J = 2.5 Hz, 1 H, H^{12}), 2.55–
2.46 (m, 2 H, H^2), 2.42 (ddd, J = 14.8, 12.5, 3.7 Hz, 1 H, H^{11a}), 2.33 (dd,
J = 25.7, 15.2 Hz, 1 H, H^{19a}), 2.17 (dt, J = 13.4, 5.6 Hz, 1 H, H^{16a}), 2.06–
1.86 (m, 5 H, $H^{1a,9,15a,18,19b}$), 1.84 (bd, J = 14.9 Hz, 1 H, H^{11b}), 1.67–1.57

(m, 4 H, H^{1b,7a,22}), 1.53 (dd, $J = 12.9, 2.0$ Hz, 1 H, H^{6b}), 1.52–1.41 (m, 5 H, H^{5,6a,27}), 1.39–1.32 (m, 2 H, H^{21}), 1.32–1.19 (m, 6 H, H^{7b,15b,16b,26}), 1.10 (s, 3 H, H^{23}), 1.04 (s, 3 H, H^{24}), 1.00 (s, 3 H, H^{29}), 0.96 (s, 3 H, H^{25}), 0.90 (s, 3 H, H^{30}). **^{13}C-NMR (150 MHz, CDCl$_3$):** δ 217.9 (C^3), 179.0 (C^{28}), 91.8 (C^{13}), 56.3 (C^{12}), 54.6 (C^5), 52.5 (C^{18}), 47.3 (C^4), 45.6 (C^{17}), 44.9 (C^9), 43.6 (C^{14}), 42.4 (C^8), 40.1 (C^{19}), 39.2 (C^1), 36.4 (C^{10}), 34.1 (C^7), 34.0 (C^{21}), 33.9 (C^2), 33.4 (C^{30}), 32.0 (C^{20}), 31.0 (C^{11}), 29.3 (C^{15}), 27.6 (C^{22}), 27.1 (C^{23}), 23.7 (C^{29}), 21.4 (C^{16}), 21.1 (C^{27}), 21.0 (C^{24}), 19.2 (C^6), 18.8 (C^{26}), 17.2 (C^{25}). **$[\alpha]_D^{20}$** +84.0 (c 1.0, CDCl$_3$). **HRMS (ESI-TOF)** m/z: $[M + Na]^+$ Calcd for C$_{30}$H$_{45}$ClNaO$_3{}^+$ 555.2444; Found 555.2439.

3.2.5 12α-Chloro-3-(hydroxyimino)-olean-28,13β-olide (6a)

A round bottom flask, equipped with magnetic stirring bar, was charged with a solution of oleanolic acid (**5**) (1.00 mmol, 457 mg, 1.00 equiv) in DCM (12 ml) and pyridine (15.2 mmol, 1229 µl, 15.2 equiv). After addition of trichloroisocyanuric acid (0.80 mmol, 186 mg, 0.80 equiv) was added, and the mixture was left stirring at 20 °C for 90 min. Then i-PrOH (0.80 mmol, 61 µl, 0.80 equiv) was added, and the mixture was left stirring at 20 °C for 30 min. Subsequently, hydroxylamine hydrochloride (3.2 mmol, 222 mg, 3.2 equiv) was added, and the reaction was left stirring at 20 °C for 19 h. The solution was diluted with DCM (8 ml), filtered over a glass frit (P 16) and the residue was washed with DCM (2 × 2 ml). The organics were washed with HCl solution (1 M in H$_2$O) (3 × 12 ml). After drying over MgSO$_4$ the solution was concentrated under reduced pressure. The residue was dried *in vacuo*

to give the target compound as pale yellow solid (0.98 mmol, 495 mg, 98 %).

^{1}H-NMR (600 MHz, CDCl$_3$): δ 7.51 (s, 1 H, H$^{3\text{-NOH}}$), 4.18 (dd, $J = 3.9, 2.3$ Hz, 1 H, H^{12}), 2.97 (ddd, $J = 15.5, 6.1, 4.2$ Hz, 1 H, H^{2a}), 2.37–2.24 (m, 2 H, H^{2b,11a}), 2.19–2.12 (m, 2 H, H^{16a,19a}), 2.02 (d, $J = 9.1$ Hz, 2 H, H^{18,19b}), 1.95 (td, $J = 13.7, 6.1$ Hz, 1 H, H^{15a}), 1.83–1.72 (m, 3 H, H^{1a,9,11b}), 1.64 (dd, $J = 9.0, 3.5$ Hz, 2 H, H^{22a}), 1.61–1.45 (m, 3 H, H^{7a,6ab}), 1.40–1.26 (m, 7 H, H^{7b,16b,21ab,27}), 1.25–1.19 (m, 5 H, H^{1b,15b,26}), 1.18–1.14 (m, 4 H, H^{5,23}), 1.06 (s, 3 H, H^{24}), 0.99 (s, 3 H, H^{29}), 0.98 (s, 3 H, H^{25}), 0.90 (s, 3 H, H^{30}). **^{13}C-NMR (150 MHz, CDCl$_3$):** δ 179.2 (C^{28}), 167.1 (C^3), 91.7 (C^{13}), 65.1 (C^{12}), 55.4 (C^5), 52.1 (C^{18}), 45.4 (C^{17}), 44.5 (C^9), 43.2 (C^{14}), 42.5 (C^8), 40.3 (C^4), 39.9 (C^{19}), 38.6 (C^1), 36.7 (C^{10}), 34.2 (C^7), 34.0 (C^{21}), 33.4 (C^{29}), 32.0 (C^{20}), 29.7 (C^{11}), 29.1 (C^{15}), 27.6 (C^{23}), 27.6 (C^{22}), 23.8 (C^{30}), 23.0 (C^{24}), 21.4 (C^{16}), 20.2 (C^{27}), 18.8 (C^{26}), 18.6 (C^6), 16.9 (C^2), 16.7 (C^{25}). $[\alpha]_D^{20}$ +27.0 (c 1.0, CDCl$_3$). **IR (neat)** 3268, 2930, 2866, 1768, 1464. **HRMS (ESI-TOF)** m/z: [M + Na]$^+$ Calcd for C$_{30}$H$_{46}$ClNNaO$_3$$^+$ 526.3058; Found 526.3059.

3.2.6 12α-Bromo-3-(hydroxyimino)-olean-28,13β-olide (6b)

A round bottom flask, equipped with magnetic stirring bar, was charged with a solution of oleanolic acid (**5**) (1.00 mmol, 457 mg, 1.00 equiv) in DCM (12 ml) and pyridine (15.2 mmol, 1229 µl, 15.2 equiv). After addition of NBS (1.05 mmol, 187 mg, 1.05 equiv), the reaction was left stirring at 20 °C for 60 min. trichloroisocyanuric acid (0.80 mmol, 186 mg, 0.80 equiv) was added subsequently, and the mixture was left stirring at 20 °C for 90 min.

Then i-PrOH (1.60 mmol, 122 µl, 1.60 equiv) was added, and the mixture was left stirring at 20 °C for 30 min. Subsequently, hydroxylamine hydrochloride (3.2 mmol, 222 mg, 3.2 equiv) was added, and the reaction was left stirring at 20 °C for 21 h. Afterwards, the solution was diluted with DCM (12 ml), filtered over a glass frit (P 16) and the residue was washed with DCM (2 × 4 ml). The organics were washed with HCl solution (1 M in H_2O) (32 ml). The aqueous phase was extracted with DCM (2 × 16 ml). The combined organic layers were washed with NaOH solution (1 M in H_2O) (32 ml), followed by NH_4Cl solution (half-saturated in H_2O) (16 ml). After drying over $MgSO_4$ the solution was concentrated under reduced pressure. The residue was dried *in vacuo* to give the target compound as pale yellow solid (0.99 mmol, 555 mg, 99 %). The obtained product was used without further purification. In a 20 mmol scale the target compound was obtained in 97 % yield. The spectroscopic data is in accordance with the literature.[5]

^{1}H-NMR (600 MHz, CDCl$_3$): δ 7.83 (s, 1 H, H$^{3\text{-NOH}}$), 4.30 (t, $J = 2.9\,\text{Hz}$, 1 H, H^{12}), 2.97 (ddd, $J = 15.5, 6.1, 4.1\,\text{Hz}$, 1 H, H^{2a}), 2.44–2.29 (m, 3 H, H^{2b,11a,19a}), 2.16 (td, $J = 13.4, 5.7\,\text{Hz}$, 1 H, H^{16a}), 2.05–1.92 (m, 3 H, H^{15a,18,19b}), 1.84 (d, $J = 15.2\,\text{Hz}$, 1 H, H^{11b}), 1.82–1.77 (m, 2 H, H^{1a,9}), 1.64 (dd, $J = 9.3, 3.5\,\text{Hz}$, 2 H, H^{22ab}), 1.61–1.44 (m, 3 H, H^{6ab,7a}), 1.43 (s, 3 H, H^{27}), 1.39–1.31 (m, 2 H, H^{21ab}), 1.31–1.19 (m, 7 H, H^{1b,7b,15b,16b,26}), 1.19–1.14 (m, 4 H, H^{5,23}), 1.06 (s, 3 H, H^{24}), 0.99 (s, 3 H, H^{29}), 0.98 (s, 3 H, H^{25}), 0.90 (s, 3 H, H^{30}). **^{13}C-NMR (150 MHz, CDCl$_3$):** δ 179.0 (C^{28}), 167.1 (C^3), 91.8 (C^{13}), 56.4 (C^{12}), 55.4 (C^5), 52.5 (C^{18}), 45.7 (C^{17}), 45.3 (C^9), 43.6 (C^{14}), 42.5 (C^8), 40.4 (C^4), 40.1 (C^{19}), 38.4 (C^1), 36.7 (C^{10}), 34.3 (C^7), 34.0 (C^{21}), 33.4 (C^{29}), 32.0 (C^{20}), 30.8 (C^{11}), 29.3 (C^{15}), 27.6 (C^{22}), 27.6 (C^{23}), 23.7 (C^{30}), 22.9 (C^{24}), 21.4 (C^{16}), 21.1 (C^{27}), 19.0 (C^{26}), 18.6 (C^6), 16.9 (C^2), 16.9 (C^{25}). **$[\alpha]_D^{20}$** +35.5 (c 1.0, CDCl$_3$). **IR (neat)** 3280, 2956, 2932, 2868, 1775, 1465. **HRMS (ESI-TOF)** m/z: $[\text{M} + \text{Na}]^+$ Calcd for $C_{30}H_{46}BrNNaO_3^+$ 570.2553; Found 570.2550.

3.2.7 12α-Chloro-3-(methoxyimino)-olean-28,13β-olide (29a)

A round bottom flask, equipped with magnetic stirring bar, was charged with a solution of **27a** (2 mmol, 978 mg, 1 equiv) in pyridine (20 ml). After addition of methoxyamine hydrochloride (4 mmol, 334 mg, 2 equiv), the flask was immersed in a preheated oilbath (80 °C), and the reaction was left stirring at 80 °C for 1.5 h. After cooling to room temperature, the mixture was diluted with DCM (200 ml). The organics were washed with HCl solution (1 M in H_2O) (3 × 125 ml) and dried over $MgSO_4$. Removal of the solvents under reduced pressure, followed by drying *in vacuo* gave the target compound as off-white foam (quantitative yield).

^{1}H-NMR (700 MHz, CDCl$_3$): δ 4.17 (dd, $J = 3.9, 2.3$ Hz, 1 H, H^{12}), 3.82 (s, 3 H, H^{31}), 2.88 (ddd, $J = 15.6, 6.1, 4.1$ Hz, 1 H, H^{2a}), 2.31–2.24 (m, 2 H, H^{2b,11a}), 2.19–2.13 (m, 2 H, H^{16a,19a}), 2.06–1.98 (m, 2 H, H^{18,19b}), 1.95 (td, $J = 13.8, 6.1$ Hz, 1 H, H^{15b}), 1.75 (m, 3 H, H^{1a,9,11b}), 1.64 (dd, $J = 9.3, 3.6$ Hz, 2 H, H^{22ab}), 1.61–1.52 (m, 2 H, H^{6a,7a}), 1.52–1.44 (m, 1 H, H^{6b}), 1.40–1.33 (m, 4 H, H^{21a,27}), 1.34–1.24 (m, 3 H, H^{7b,16b,21b}), 1.23 (s, 5 H, H^{1b,15b,26}), 1.16 (s, 3 H, H^{23}), 1.14 (dd, $J = 11.9, 2.2$ Hz, 1 H, H^5), 1.05 (s, 3 H, H^{24}), 0.99 (s, 3 H, H^{29}), 0.96 (s, 3 H, H^{25}), 0.90 (s, 3 H, H^{30}).

^{13}C-NMR (175 MHz, CDCl$_3$): δ 179.2 (C^{28}), 165.7 (C^3), 91.8 (C^{13}), 65.1 (C^{12}), 61.2 (C^{31}), 55.5 (C^5), 52.1 (C^{18}), 45.5 (C^{17}), 44.5 (C^9), 43.2 (C^{14}), 42.5 (C^8), 40.1 (C^4), 39.9 (C^{19}), 38.7 (C^1), 36.6 (C^{10}), 34.2 (C^7), 34.1 (C^{21}), 33.4 (C^{29}), 32.0 (C^{20}), 29.7 (C^{11}), 29.1 (C^{15}), 27.7 (C^{23}), 27.6 (C^{22}), 23.8 (C^{30}), 23.1 (C^{24}), 21.4 (C^{16}), 20.2 (C^{27}), 18.8 (C^{26}), 18.6 (C^6), 17.7

(C^2), 16.7 (C^{25}). $[\alpha]_D^{20}$ +39.0 (c 1.0, CDCl$_3$). **HRMS (ESI-TOF)** m/z: [M + Na]$^+$ Calcd for $C_{31}H_{48}ClNNaO_3^+$ 540.3215; Found 540.3212.

3.2.8 23-Acetoxy-3-(acetoxyimino)-12α-chloro-olean-28,13β-olide (7a)

A screw cap vial, equipped with magnetic stirring bar, was charged with a suspension of **6a** (100 µmol, 50.4 mg, 1 equiv) in a mixture of Ac$_2$O (2.65 mmol, 250 µl, 26.5 equiv) and AcOH (250 µl). The mixture was left stirring at 20 °C for 5 h. Pd(OAc)$_2$ (12.5 µmol, 2.8 mg, 0.125 equiv) and PhI(OAc)$_2$ (150 µmol, 48.3 mg, 1.5 equiv) were added subsequently and the reaction was left stirring at 20 °C for 48 h. Solvents were removed under reduced pressure at 50 °C. Purification by column chromatography (silica gel, heptane/EtOAc = 4:1) gave the target compound as colourless solid (34.4 µmol, 20.8 mg, 34 %; *dr* 80:20).

^{1}H-NMR (600 MHz, CDCl$_3$, major isomer): δ 4.22 (d, $J = 11.0$ Hz, 1 H, H^{23a}), 4.21–4.16 (m, 1 H, H^{12}), 4.10 (d, $J = 11.0$ Hz, 1 H, H^{23b}), 2.74–2.63 (m, 1 H, H^{2a}), 2.34–2.25 (m, 2 H, H^{2b,11a}), 2.21–2.12 (m, 5 H, H^{16a,19a,32}), 2.07 (s, 3 H, H^{34}), 2.05–2.00 (m, 2 H, H^{18,19b}), 1.94 (td, $J = 13.8, 6.1$ Hz, 1 H, H^{15a}), 1.84–1.77 (m, 2 H, H^{1a,9}), 1.76–1.70 (m, 1 H, H^{11b}), 1.68–1.62 (m, 2 H, H^{22ab}), 1.57–1.48 (m, 3 H, H^{5,6a,7a}), 1.47–1.41 (m, 1 H, H^{6b}), 1.39 (s, 3 H, H^{27}), 1.36–1.26 (m, 5 H, H^{1b,7b,16b,21ab}), 1.26–1.21 (m, 4 H, H^{15b,26}), 1.19 (s, 3 H, H^{24}), 0.99 (s, 3 H, H^{29}), 0.94 (s, 3 H, H^{25}), 0.91 (s, 3 H, H^{30}). **^{13}C-NMR (150 MHz, CDCl$_3$, major isomer):** δ 179.0 (C^{28}), 171.1 (C^{33}), 170.4 (C^{31}), 169.7 (C^3), 91.6 (C^{13}), 68.4 (C^{23}), 65.0

(C^{12}), 52.1 (C^{18}), 48.3 (C^5), 45.4 (C^{17}), 44.1 (C^4), 44.1 (C^9), 43.2 (C^{14}), 42.2 (C^8), 39.9 (C^{19}), 37.4 (C^1), 36.2 (C^{10}), 34.0 (C^{21}), 33.7 (C^7), 33.4 (C^{29}), 32.0 (C^{20}), 29.9 (C^{11}), 29.0 (C^{15}), 27.6 (C^{22}), 23.7 (C^{30}), 21.4 (C^{16}), 21.2 (C^{34}), 20.4 (C^2), 20.2 (C^{32}), 20.1 (C^{27}), 19.0 (C^6), 18.9 (C^{24}), 18.6 (C^{26}), 17.1 (C^{25}). **IR (neat)** 2941, 2869, 1771, 1743, 1466. **HRMS (ESI-TOF)** m/z: $[M + Na]^+$ Calcd for $C_{34}H_{50}ClNNaO_6^+$ 626.3219; Found 626.3208.

3.2.9 23-Acetoxy-3-(acetoxyimino)-12α-bromo-olean-28,13β-olide (7b)

A round bottom flask, equipped with magnetic stirring bar, was charged with a suspension of **6b** (19.7 mmol, 10.805 g, 1 equiv) in a mixture of Ac_2O (512 mmol, 48.4 ml, 26 equiv) and AcOH (49.25 ml). The flask was immersed into a preheated oilbath (40 °C), and the mixture was left stirring at 40 °C for 2 h. $Pd(OAc)_2$ (2.95 mmol, 663.0 mg, 0.15 equiv) and $PhI(OAc)_2$ (31.5 mmol, 10.150 g, 1.6 equiv) were added subsequently and the reaction was left stirring at 40 °C for 16 h. After cooling to room temperature, MeOH (49.25 ml) was added and the resulting mixture was left stirring at room temperature for 30 min. Solvents were removed under reduced pressure at 50 °C. Purification by column chromatography (silica gel, heptane/EtOAc = 4:1) gave the target compound as colourless solid (9.4 mmol, 6.100 g, 48 %; *dr* 75:25).
1H-NMR (600 MHz, CDCl₃, major): δ 4.31 (dd, $J = 3.9, 2.3$ Hz, 1 H, H^{12}), 4.22 (d, $J = 11.0$ Hz, 1 H, H^{23a}), 4.10 (d, $J = 11.0$ Hz, 1 H, H^{23b}), 2.74–2.63 (m, 1 H, H^{2a}), 2.46–2.35 (m, 1 H, H^{11a}), 2.34–2.26 (m, 2 H, $H^{2b,19a}$), 2.20–2.13 (m, 4 H, $H^{16a,32}$), 2.08 (s, 3 H, H^{34}), 2.01–1.98 (m, 2 H, $H^{18,19b}$),

1.95 (td, J = 13.5, 5.6 Hz, 1 H, H^{15a}), 1.87–1.72 (m, 3 H, H^{1a,9,11a}), 1.67–1.62 (m, 2 H, H^{22ab}), 1.56–1.49 (m, 3 H, H^{5,6a,7a}), 1.46–1.42 (m, 4 H, H^{6b,27}), 1.40–1.32 (m, 3 H, H^{1b,21ab}), 1.31–1.20 (m, 6 H, H^{7b,15b,16b,26}), 1.19 (s, 3 H, H^{24}), 1.00 (s, 3 H, H^{29}), 0.94 (s, 3 H, H^{25}), 0.90 (s, 3 H, H^{30}). **^{13}C-NMR (150 MHz, CDCl$_3$, major):** δ 178.9 (C^{28}), 171.1 (C^{33}), 170.4 (C^{31}), 169.7 (C^3), 91.6 (C^{13}), 68.4 (C^{23}), 56.3 (C^{12}), 52.5 (C^{18}), 48.3 (C^5), 45.6 (C^9), 44.9 (C^{17}), 44.1 (C^4), 43.6 (C^{14}), 42.3 (C^8), 40.1 (C^{19}), 37.3 (C^1), 36.2 (C^{10}), 34.0 (C^{21}), 33.8 (C^7), 33.4 (C^{29}), 32.0 (C^{20}), 31.0 (C^{11}), 29.2 (C^{15}), 27.6 (C^{22}), 23.7 (C^{30}), 21.4 (C^{16}), 21.2 (C^{27}), 21.0 (C^{34}), 20.3 (C^2), 20.1 (C^{32}), 19.0 (C^6), 18.9 (C^{26}), 18.8 (C^{24}), 17.3 (C^{25}). **IR (neat)** 2928, 2863, 1770, 1743, 1465. **HRMS (ESI-TOF)** m/z: [M + Na]$^+$ Calcd for C$_{34}$H$_{50}$BrNNaO$_6$$^+$ 670.2714; Found 670.2712.

3.2.10 12α-Bromo-23-hydroxy-3-oxo-olean-28,13β-olide (8b)

A round bottom flask, equipped with magnetic stirring bar, was charged with a solution of **7b** (9.29 mmol, 6.026 g, 1 equiv) and K$_2$CO$_3$ (9.29 mmol, 1.284 g, 1 equiv) in MeOH (93 ml). The flask was immersed into a preheated oilbath (60 °C) and the reaction was left stirring at 60 °C for 1 h. After addition of THF (93 ml) and a solution of copper sulfate pentahydrate (46.4 mmol, 11.597 g, 5 equiv) in H$_2$O (93 ml), the resulting suspension was left stirring at 60 °C for 16 h. After cooling to room temperature, the mixture was extracted with EtOAc (1 × 465 ml and 2 × 232 ml). The combined organic phases were dried over MgSO$_4$ and concentrated under reduced pressure. Purification by column chromatography (silica gel, heptane/EtOAc = 1:2) gave the target compound as colourless solid (3.83 mmol, 2.105 g, 55 %; *dr* > 20:1).

^{1}H-NMR (700 MHz, CDCl$_3$): δ 4.31 (dd, $J = 3.9, 2.3$ Hz, 1 H, H^{12}), 3.69 (dd, $J = 11.2, 6.4$ Hz, 1 H, H^{23a}), 3.43 (dd, $J = 11.3, 6.9$ Hz, 1 H, H^{23b}), 2.64 (ddd, $J = 16.4, 12.4, 7.2$ Hz, 1 H, H^{2a}), 2.45 (ddd, $J = 14.9, 12.3, 3.9$ Hz, 1 H, H^{11a}), 2.37 (ddd, $J = 16.4, 6.1, 2.9$ Hz, 1 H, H^{2b}), 2.33 (dd, $J = 10.1, 1.9$ Hz, 1 H, H^{19a}), 2.23 (t, 1 H, H$^{23\text{-OH}}$), 2.17 (td, $J = 13.4, 5.8$ Hz, 1 H, H^{16a}), 2.05–1.94 (m, 4 H, H^{1a,15a,18,19b}), 1.92 (dd, 1 H, H^9), 1.88 (dt, 1 H, H^{11b}), 1.77 (dd, $J = 12.1, 2.4$ Hz, 1 H, H^5), 1.67–1.63 (m, 3 H, H^{7a,22ab}), 1.56 (s, 2 H, H^{1b,6a}), 1.46 (s, 3 H, H^{27}), 1.42 (dq, $J = 13.3, 3.2$ Hz, 1 H, H^{6b}), 1.39–1.32 (m, 2 H, H^{21ab}), 1.32–1.21 (m, 6 H, H^{7b,15b,16b,26}), 1.09 (s, 3 H, H^{25}), 1.01 (s, 3 H, H^{24}), 1.00 (s, 3 H, H^{29}), 0.91 (s, 3 H, H^{30}). **^{13}C-NMR (175 MHz, CDCl$_3$):** δ 218.5 (C^3), 178.9 (C^{28}), 91.7 (C^{13}), 67.0 (C^{23}), 56.2 (C^{12}), 52.6 (C^4), 52.5 (C^{18}), 48.8 (C^5), 45.7 (C^{17}), 45.0 (C^9), 43.7 (C^{14}), 42.5 (C^8), 40.1 (C^{19}), 38.8 (C^1), 36.3 (C^{10}), 35.3 (C^2), 34.1 (C^7), 34.0 (C^{21}), 33.4 (C^{29}), 32.0 (C^2), 30.9 (C^{11}), 29.3 (C^{15}), 27.6 (C^{22}), 23.7 (C^{30}), 21.4 (C^{16}), 21.2 (C^{27}), 19.1 (C^{26}), 18.7 (C^6), 16.8 (C^{25}), 16.8 (C^{24}). $[\alpha]_D^{20}$ +53.4 (*c* 1.0, CDCl$_3$); **IR (neat)** 3486, 2937, 2869, 1772, 1698, 1461. **HRMS (ESI-TOF)** m/z: [M + Na]$^+$ Calcd for C$_{30}$H$_{45}$BrNaO$_4^+$ 571.2393; Found 571.2394.

3.2.11 Hederagonic acid
(23-Hydroxy-3-oxo-olean-12-en-28-oic acid) (1)

A round bottom flask, equipped with magnetic stirring bar, was charged with a solution of **8b** (36.4 µmol, 20.0 mg, 1 equiv) and zinc dust (1092 µmol, 71.4 mg, 30 equiv) in AcOH (360 µl). The flask was immersed into a preheated oilbath (40 °C) and the reaction was left stirring at 40 °C for 2 h. After cooling to room temperature, the mixture was diluted with EtOAc (730 µl), filtered over a silica pad and was the residue was washed with EtOAc (2 × 360 µl). The solution was concentrated under reduced pressure. Purification by column chromatography (silica gel, heptane/EtOAc/AcOH = 55:40:5) gave the target compound as colourless solid (34.6 µmol, 16.3 mg, 95 %). 20 % overall yield starting from **5**. The spectroscopic data is in accordance with the literature.[1,63]

^{1}H-NMR (600 MHz, CDCl$_3$): δ 5.31 (t, $J = 3.7$ Hz, 1 H, H^{12}), 3.65 (d, $J = 11.3$ Hz, 1 H, H^{23a}), 3.42 (d, $J = 11.4$ Hz, 1 H, H^{23b}), 2.83 (dd, $J = 13.9, 4.7$ Hz, 1 H, H^{18}), 2.63 (ddd, $J = 16.1, 13.3, 6.8$ Hz, 1 H, H^{2a}), 2.27 (ddd, $J = 16.2, 5.3, 2.5$ Hz, 1 H, H^{2b}), 2.04–1.95 (m, 2 H, H^{11a,16a}), 1.95–1.89 (m, 2 H, H^{1a,11b}), 1.77 (td, $J = 13.9, 4.5$ Hz, 1 H, H^{7a,22ab}), 1.74–1.67 (m, 2 H, H^{9,22a}), 1.66–1.54 (m, 4 H, H^{5,7b,16b,19a}), 1.54–1.46 (m, 2 H, H^{6a,15a}), 1.43–1.32 (m, 4 H, H^{1b,6b,15b,21a}), 1.27–1.20 (m, 1 H, H^{21b}), 1.19–1.13 (m, 7 H, H^{19b,25,27}), 1.13–1.08 (m, 1 H, H^{22b}), 1.01 (s, 3 H, H^{24}), 0.93 (s, 3 H, H^{30}), 0.90 (s, 3 H, H^{29}), 0.83 (s, 3 H, H^{26}). **^{13}C-NMR (150 MHz, CDCl$_3$):** δ 219.3 (C^3), 182.1 (C^{28}), 143.9 (C^{13}), 122.4 (C^{12}), 67.0 (C^{23}), 52.5 (C^4), 49.3 (C^5), 46.9 (C^9), 46.6 (C^{17}), 45.9 (C^{19}), 41.9 (C^{14}), 41.2 (C^{18}), 39.4 (C^8), 38.9 (C^1), 36.8 (C^{10}), 35.3 (C^2), 33.9 (C^{21}), 33.2 (C^{29}), 32.5 (C^7), 32.2 (C^{15}), 30.8 (C^{20}), 27.8 (C^{22}), 26.0 (C^{27}), 23.7 (C^{30}), 23.6 (C^{11}), 23.1 (C^{16}), 19.2 (C^6), 17.3 (C^{26}), 17.0 (C^{24}), 15.3 (C^{25}). **[α]$_D^{20}$** +70.9 (c 1.0, CDCl$_3$). **HRMS (ESI-TOF)** m/z: [M + Na]$^+$ Calcd for C$_{30}$H$_{45}$BrNaO$_4$$^+$ 493.3288; Found 493.3288.

References and Notes

[1] X.-A. Wen, J. Liu, L.-Y. Zhang, P.-Z. Ni, H.-B. Sun, *Chin. J. Nat. Med.* **2010**, *8*, 441–448.

[2] L. V. Desai, K. L. Hull, M. S. Sanford, *J. Am. Chem. Soc.* **2004**, *126*, 9542–9543.

[3] A. R. Dick, K. L. Hull, M. S. Sanford, *J. Am. Chem. Soc.* **2004**, *126*, 2300–2301.

[4] I. Dip, C. Gethers, T. Rice, T. S. Straub, *Tetrahedron Lett.* **2017**, *58*, 2720–2722.

[5] D. Kaminskyy, B. Bednarczyk-Cwynar, O. Vasylenko, O. Kazakova, B. Zimenkovsky, L. Zaprutko, R. Lesyk, *Med. Chem. Res.* **2011**, *21*, 3568–3580.

[6] B.-X.-Z. Liu, J.-Y. Zhou, Y. Li, X. Zou, J. Wu, J.-F. Gu, J.-R. Yuan, B.-J. Zhao, L. Feng, X.-B. Jia, R.-P. Wang, *BMC Complement. Altern. Med.* **2014**, *14*, 412.

[7] W. A. Jacobs, *J. Biol. Chem.* **1925**, *63*, 631–640.

[8] S. K. Agarwal, R. P. Rastogi, *Phytochemistry* **1974**, *13*, 666–668.

[9] N. K. Hart, J. A. Lamberton, A. A. Sioumis, H. Suares, *Aust. J. Chem.* **1976**, *29*, 655.

[10] K. Jatczak, G. Grynkiewicz, *Acta Biochim Pol* **2014**, *61*, 227–243.

[11] H. Yao, J. Duan, J. Wang, Y. Li, *Biochem. Syst. Ecol.* **2012**, *42*, 14–17.

[12] M. Miyakoshi, K. Shirasuna, Y. Hirai, K. Shingu, S. Isoda, J. Shoji, Y. Ida, T. Shimizu, *J. Nat. Prod.* **1999**, *62*, 445–448.

[13] M. D. Greca, A. Fiorentino, P. Monaco, L. Previtera, *Phytochemistry* **1993**, *35*, 201–204.

[14] European and Mediterranean Plant Protection Organization, *Bull. OEPP* **2014**, *44*, 457–471.

[15] J.-J. Li, J. Yang, F. Lü, Y.-T. Qi, Y.-Q. Liu, Y. Sun, Q. Wang, *Chin. J. Nat. Med.* **2012**, *10*, 279–283.

[16] C. H. Greenberg, L. M. Smith, D. J. Levey, *Biol. Invasions* **2001**, *3*, 363–372.

[17] C.-S. Chang, H. Kim, H.-S. Kang, D. K. Lee, *Bot. Bull. Acad. Sin.* **2003**, *44*, 337–344.

[18] Z. Shu, Z. Chen, X.-j. Ding, B.-q. Lu, C.-j. Ji, Q.-m. X. X.-r. Li, S.-l. Yang, *Heterocycles* **2011**, *83*, 2365.

[19] Y. Xu, L. Bai, Y. Liu, Y. Liu, T. Xu, S. Xie, Y. Si, H. Zhou, T. Liu, D. Xu, *Molecules* **2010**, *15*, 1891–1897.

[20] H. Duan, Y. Takaishi, H. Momota, Y. Ohmoto, T. Taki, Y. Jia, D. Li, *Phytochemistry* **2000**, *53*, 805–810.

[21] (a) R. Vidéki, Floating pennywort, Hydrocotyle ranunculoides (Apiales: Apiaceae) - 5396697, Doronicum Kft., **2009**, Bugwood.org (visited on 2020-02-24); (b) J. R. Allison, Oriental bittersweet, Celastrus orbiculatus (Celastrales: Celastraceae) - 2105096, Georgia Department of Natural Resources, **2010**, http://Bugwood.org (visited on 2020-02-24).

[22] Z. Liang, L. Zhang, L. Li, J. Liu, H. Li, L. Zhang, L. Chen, K. Cheng, M. Zheng, X. Wen, P. Zhang, J. Hao, Y. Gong, X. Zhang, X. Zhu, J. Chen, H. Liu, H. Jiang, C. Luo, H. Sun, *Eur. J. Med. Chem.* **2011**, *46*, 2011–2021.

[23] J. Pollier, A. Goossens, *Phytochemistry* **2012**, *77*, 10–15.

[24] J. Chen, J. Liu, L. Zhang, G. Wu, W. Hua, X. Wu, H. Sun, *Bioorg. Med. Chem. Lett.* **2006**, *16*, 2915–2919.

[25] J. E. Baldwin, C. Nájera, M. Yus, *J. Chem. Soc. Chem. Commun.* **1985**, 126–127.

[26] A. G. Constable, W. S. McDonald, L. C. Sawkins, B. L. Shaw, *J. Chem. Soc. Dalton Trans.* **1980**, 1992–2000.

[27] K. Carr, H. M. Saxton, J. K. Sutherland, *J. Chem. Soc. Perkin Trans. 1* **1988**, 1599–1601.

[28] T. M. Peakman, H. L. ten Haven, J. Rullkötter, J. A. Curiale, *Tetrahedron* **1991**, *47*, 3779–3786.

[29] L. Bore, T. Honda, G. W. Gribble, *J. Org. Chem.* **2000**, *65*, 6278–6282.

[30] B. V. S. Reddy, L. R. Reddy, E. J. Corey, *Org. Lett.* **2006**, *8*, 3391–3394.

[31] V. V. Dunina, O. A. Zalevskaya, V. M. Potapov, *Russ. Chem. Rev.* **1988**, *57*, 250–269.

[32] S. R. Neufeldt, M. S. Sanford, *Acc. Chem. Res.* **2012**, *45*, 936–946.

[33] M. Berger, 'PhD Thesis, in progress', Maulide Group, University of Vienna, Vienna, Austria, **2017**.

[34] J. A. Kerr, *Chem. Rev.* **1966**, *66*, 465–500.

[35] S. R. Neufeldt, M. S. Sanford, *Org. Lett.* **2010**, *12*, 532–535.

[36] K. G. Lewis, D. J. Tucker, *Aust. J. Chem.* **1983**, *36*, 2297.

[37] Y. Hayashi, *Chem. Sci.* **2016**, *7*, 866–880.

[38] P. A. Wender, *Nat. Prod. Rep.* **2014**, *31*, 433–440.

[39] E. J. Corey, G. Schmidt, *Tetrahedron Lett.* **1979**, *20*, 399–402.

[40] E. J. Corey, J. W. Suggs, *Tetrahedron Lett.* **1975**, *16*, 2647–2650.

[41] B. E. Rivero-Chan, J. G. Marrero, S. Hernández-Ortega, G. J. Mena-Rejón, L. D. Miranda, *Nat. Prod. Res.* **2012**, *26*, 675–679.

[42] U. Tilstam, H. Weinmann, *Org. Process Res. Dev.* **2002**, *6*, 384–393.

[43] E. C. Juenge, P. L. Spangler, W. P. Duncan, *J. Org. Chem.* **1966**, *31*, 3836–3838.

[44] T. R. Walters, W. W. Zajac, J. M. Woods, *J. Org. Chem.* **1991**, *56*, 316–321.

[45] H. E. Elsayed, M. R. Akl, H. Y. Ebrahim, A. A. Sallam, E. G. Haggag, A. M. Kamal, K. A. E. Sayed, *Chem. Biol. Drug Des.* **2014**, *85*, 231–243.

[46] Y. Y. See, A. T. Herrmann, Y. Aihara, P. S. Baran, *J. Am. Chem. Soc.* **2015**, *137*, 13776–13779.

[47] D. H. R. Barton, P. de Mayo, *J. Chem. Soc. (Resumed)* **1954**, 887.

[48] S. R. Johns, J. A. Lamberton, T. C. Morton, H. Suares, R. I. Willing, *Aust. J. Chem.* **1983**, *36*, C-4-epi hederagonic acid, 2537.

[49] R. G. Soengas, H. Rodríguez-Solla, A. Díaz-Pardo, R. Acúrcio, C. Concellón, V. del Amo, A. M. S. Silva, *Eur. J. Org. Chem.* **2015**, *2015*, 2524–2530.

[50] M. B. Sporn, K. T. Liby, M. M. Yore, L. Fu, J. M. Lopchuk, G. W. Gribble, *J. Nat. Prod.* **2011**, *74*, 537–545.

[51] Chemical Abstracts Service: Columbus, OH, Ed., SciFinder, **2017**, https://scifinder.cas.org (visited on 2017-10-23).

[52] abcr GmbH, abcr Webshop, Article number: AB251630, https:// www.abcr.de/shop/de/Oleanolicacid-hydrate-95-327970.html/ (visited on 2017-12-13).

[53] Molport, Molport, Article number: MolPort-035-706-388, https://www.molport.com/shop/moleculelink/WCXZTKJFWJFMJG-AOIBKUJJSA-N/35706388 (visited on 2017-12-13).

[54] G. R. Fulmer, A. J. M. Miller, N. H. Sherden, H. E. Gottlieb, A. Nudelman, B. M. Stoltz, J. E. Bercaw, K. I. Goldberg, *Organometallics* **2010**, *29*, 2176–2179.

[55] Bruker AXS Inc., Madison, Wisconsin, USA, SAINT, **2012**.

[56] Bruker AXS Inc., Madison, Wisconsin, USA, SADABS, **2001**.

[57] O. V. Dolomanov, L. J. Bourhis, R. J. Gildea, J. A. K. Howard, H. Puschmann, *J. Appl. Crystallogr.* **2009**, *42*, 339–341.

[58] G. M. Sheldrick, *Acta Crystallogr. C* **2015**, *71*, 3–8.

[59] G. M. Sheldrick, *Acta Crystallogr. A* **2008**, *64*, 112–122.

[60] A. L. Spek, *Acta Crystallogr. D* **2009**, *65*, 148–155.

[61] C. F. Macrae, P. R. Edgington, P. McCabe, E. Pidcock, G. P. Shields, R. Taylor, M. Towler, J. van de Streek, *J. Appl. Crystallogr.* **2006**, *39*, 453–457.

[62] A. Martinez, A. Perojil, F. Rivas, M. Medina-O'Donnell, A. Parra, *Tetrahedron* **2015**, *71*, Bromolactone of OA (Cmpd 2) full NMR charaterisation., 792–800.

[63] P. Bhandari, R. P. Rastogi, *Phytochemistry* **1984**, *23*, 2082–2085.

Printed in the United States
By Bookmasters